The Steam Engine in Industry
2 Mining and the Metal Trades

Historic Industrial Scenes

THE STEAM ENGINE IN INDUSTRY -2

Mining and the Metal Trades

George Watkins

MOORLAND PUBLISHING

UNIVERSITY OF
STRATHCLYDE LIBRARIES

British Library Cataloguing in Publication Data

The steam engine in industry. - (Historic
industrial scenes).
Vol.2
1. Steam-engines - Great Britain - History
- Pictorial works
I. Watkins, George II. Series
621.1'6'0941

ISBN 0-903485-66-4

ISBN 0 903485 65 6 (volume 1)
0 903485 66 4 (volume 2)
0 903485 67 2 (set of two volumes)

© G. Watkins 1979

All rights reserved. No part of this
publication may be reproduced, stored
in a retrieval system, or transmitted
in any form or by any means, electronic,
mechanical, photocopying, recording or
otherwise, without the prior permission
of Moorland Publishing Company.

Photoset by Advertiser Printers Ltd, Newton Abbot
and printed in Great Britain
by Redwood Burn Ltd, Trowbridge & Esher
for Moorland Publishing Co Ltd,
PO Box 2, Ashbourne, Derbyshire, DE6 1DZ

Contents

	page
Introduction	6
Technical Notes	8
Mine Winding Engines	12
Mine Pumping Engines	48
Mine Ventilation Engines	63
Drift Mine and Haulage Engines	69
Blast Furnace Blowing Engines	73
Rolling Mill Engines	86
Engines for Metal Working—General Processes	110
Works Services	114
Boilers	118
List of Engine Builders	124
List of Engines	125

Acknowledgements

Again, my book is a tribute to the engine owners, the managers and the enginewrights at the collieries, and iron and steel works, and the fitters and drivers everywhere. The National Coal Board was most helpful, even in the chaos of reorganisation to 1965, readily aiding the recording of engines which had to be replaced and are now unknown. I have grateful memories of many devoted men who were happy in, and proud of, the calling that was their way of life, often telling of the rapid restoration of production after seemingly hopeless breakdowns, quickly assessing the best method of repair and unstinted application of it. Again too Tony and Jane Woolrich have been invaluable in offering quiet for study and thinking, and typing the manuscript. The prints again are a tribute to Colin Wilson's art, and the sketches are by Frank Wightman. I must stress that most of my examples have disappeared and my visits were often a race with the scrappers who had their job to do. I am also grateful to the Visiting Fellowship at Bath University which maintains my involvement in technology.

Introduction

The steam engine aided many of the developments of the nineteenth century by an ample fuel supply to steam engines that gave power wherever it was required. In fact, the improvement in public services stemmed from those made in the mining and metal trades which provided the vast amounts of material used to improve the towns. The demands were parallel, all calling for ever more power, so that from the 20-25hp engines that developed industry early in the nineteenth century, by its end up to 3,000hp was used for winding, 1,000 for pumping, and 8-10,000hp as the steel trade evolved. Control of these great powers was achieved by simple mechanical linkages and servo-operated systems to relieve the operators. With the ever increasing efficiency of the iron and steel trades there was less waste or surplus heat, as more was utilised within the processes. With ever increasing power demands, economical engines became essential and this also applied to the collieries where pumping and haulage requirements became a serious drain. A large colliery would require up to twenty Lancashire boilers, two large winders, and turbine compressors and generators of 5-6000hp. The great ironworks employing 3,000 men on a dozen blast furnaces, as at Barrow, used sixty boilers in the iron works, and over a hundred in the steel works, and the great American plants had impressive arrays of a hundred or more of their smaller boilers, and twenty or more blowing engines. Even in the USA where output was the key factor, it was felt that fuel must be saved so that compounding and condensing, little known in the ferrous trades until then, developed around the turn of the century.

Early in the twentieth century, water-tube boilers, high pressures and turbines greatly reduced the heat consumption in the plants. The integrated works, in which iron was run from their own furnaces into ladles, stored in mixers, converted into steel and rolled into industrial 'uses', without re-melting, also gained high economy. The great steam engines of the nineteenth century were, then, the product of the parallel growth of facilities and concepts. Mining with its smaller groupings did not often have the impressiveness of the iron trades, but it did have its share. Probably the largest shaft pumping unit was the 130in Bull engine for the St Barbe Colliery, Belgium, and there was the twin 104in Bull engines for the Tyne Coal Co and a few 100in Cornish pumps. There were also the great Davey engines for the South Staffordshire mines drainage scheme of the 1880s, and the Lindal Iron Mines. For winding, the 68in x 7ft vertical engine for Wearmouth and three others at Ryhope were splendid machines that worked for eighty years or so; so was the 48in x 7ft Grange Ironworks twin-cylinder engine for Silksworth. In Wales was the 54in inverted vertical engine for Harris Navigation Colliery, and great Markham engines for the Yorkshire pits. The northern engines achieved high performances on modest steam specifications, often from simple egg-ended boilers and poor fuel. Their small evaporative capacity led to the use of great batteries of such boilers, and to reduce their number, especially in ironworks, they were made up to 60 or 80ft long. Simple, often roughly made, they did well on poor fuel and bad water, but they were also treacherous, creating impressive havoc when large batteries ran amok in explosions. This also applied to the many Rastrick flue boilers that drove the wrought-iron works from furnace waste heat, but their record, considering the vast numbers used, despite many explosions and lives lost, was not a bad one.

Rapid movement of the mineral was of vital importance. Underground haulage was along ever extending workings and uneven gradients, often with main roads of miles in length. On the surface, movement to the screens and picking belts, and return of the empty tubs to the pit bank and working face for re-filling, were equally important.

The main use of steam in this field was for the heavy haulage engines, usually near to the pit bottom, which pulled long trains of full tubs along the main haulage roads and returned empty ones. The great deep collieries of North East England posed many problems in haulage. Wearmouth Colliery was an instance in which, after nearly half a century, the workings were extended to 3½ miles from the shaft bottom. Handling 90 to 100 tons of coal per hour involved the use of three haulage engines underground, and over 20 miles of wire rope. A large colliery in the Midlands using horses only would require eighty to a hundred underground.

Boilers were often placed underground to drive the engines, but in later years steam was supplied by pipes from the surface. Main-road haulage engines in extensive pits were as large as many small colliery winding engines. Rope speeds were low to suit the great weights hauled, the variable gradients and the nature of the track. This allowed the use of a small engine driving through single or double reduction gearing to move great weights over long distances, and the small size of the individual parts was very valuable when lowering into, and installing in, the pit. To avoid sending steam down the pit, the engine was sometimes placed on the surface, and the ropes were passed through the shaft. An early example of this was the Monkwearmouth Colliery in the 1850s. A 120hp engine on the surface drove two ropes down the 1,700ft shaft, which each hauled 12 tons of coal up a plane 1,400 yards long. Many others followed; steam was in fact essential with steep gradients in dipping seams.

Haulage was also very important for sending coal from inland pits to the ships on the Tyne. The restricted wayleaves available meant that the coal railways had often to pass over, rather than around the hill on route, requiring the use of fixed engines and cable haulage systems. The Pinxton No 1 pit in Nottinghamshire was an early example of underground steam haulage. The engine was built in 1856 and installed about 100 yards from the shaft bottom, together with two boilers for $40lb/in^2$, when the increased dip in the main road became too great for horse haulage. It hauled fourteen trams of coal by a single rope, and the exhaust steam and smoke were discharged into the ventilating shaft. Beam engines for haulage and winding were used underground before this in Cornwall.

Capstans or haulages were also invaluable maids of all work on the pit top. They wound the spoil to the tops of waste tips, often up to a mile away and over 100ft high. They pulled full and empty trucks around the yard, and heavy duty capstans were indispensible for changing the winding ropes. In the pit special ones designed to break down into small pieces could be taken into by-roads and re-erected away from the main haulage roads. They were compact and gave a massive pull at low speed.

The rapid extension of the workings that occurred in the collieries of the North East soon led to the need for increased ventilation for safety. For many years the tendency for a column of heated air to rise in the upcast shaft served, and was aided by a large furnace discharging into it. Eppleton Colliery Co, Durham, was a good example of this. The main furnace grate was 60ft long x 11ft wide, which, using 24 tons of coal daily, and aided by the flue gases from several underground boilers, provided a flow of over 300,000 cu ft per minute. Coursing the air around the worked out areas meant long circuits, and at one pit there were 30 miles of road to traverse between the downcast and upcast shafts, some 900 yards apart. Fans developed rapidly from about 1850 and were able to provide large flow with no explosion risk and far less fuel consumption. The Waddell open runner was the largest and the most spectacular; for the most part fans were silent and unseen, and worked by enclosed Belliss type engines they ran non-stop for many months with little attention. The exception to this practice was the coals of south Staffordshire, which were so fiery that rapid ventilation set them on fire. The low air flow restricted the area worked from one shaft, so a small winder was placed to work out an area from numerous shafts. Hundreds of small Newcomen engines tended by boys were used for this, the low-cost engines being abandoned when the coal was worked out from the several shafts. By the end of the century up to 6,300 tons could be wound from a 1,500ft-deep shaft in five shifts of 8½ hours. The Cornish mining scene was similarly graced with many engines, but this was due to the small areas of valuable metal veins that each mine or 'sett' worked. We hear much of the avaricious coal owners of the past, but the fact is that without them there would have been little coal mining. Vast sums were courageously spent on difficult winnings, sometimes over years of effort. An example was the great Wallsend Colliery which started sinking in 1778; the shaft was lost owing to quicksand and water, and the colliery was only opened in 1781 after great difficulty. At Wearmouth, Pembertons began sinking in 1826, lost that shaft from sand in 1831, and nearly lost the one that followed. It was not until 1834, after spending £100,000 that they reached the Hutton coal. Murton was even worse; sinking was started in 1838, and to reach the Hutton seam the cost was over £250,000.

Nearly 160 Newcomen engines were installed in the Northern coalfield, in the 60 years after 1715, costing anything up to £1,200 each, and around £400 a year to run; the Jesmond and Byker pits used five. Such engines pumped for a quarter of the cost of horses, raising up to 250,000 gallons per day. Many small pumps were installed underground, such as the Evans type installed at Crumlin Navigation pit in South Wales. Working on steam at $80lb/in^2$ it pumped 2½ million gallons per day to 600ft. Another small Evans steam pump installed in a dip ran for six weeks under water when a flood broke in, and pumped the dip dry.

Technical Notes

Winding Drums

When raising mineral each wind started with the payload and the weight of the rope against the engine and was only offset by the weight of the downgoing cage. The load varied throughout the wind, and several methods were adopted to counteract this. The simplest was the tail rope coupled to the two cages. With this the load was reduced to the mineral payload. The great single cylinder winders of the North East used the counterbalance system (Plates 8 and 9). Conical drums which began each wind at a small radius greatly eased the starting strain on the engine, as did the diabolo drum. The flat rope, a broad tape four to six thicknesses in width and wrapped coil upon coil within the horns or side guides again started the load on the least radius. Each had its own characteristics, but I have no space to explain this.

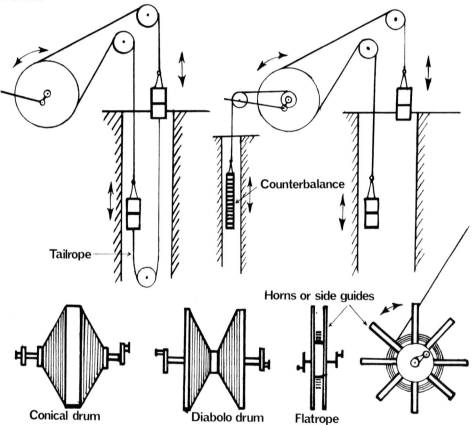

Blast Furnace Blowing Engines

At first direct coupled beam engines were used, but rotative beam blowers soon followed and they reached huge dimensions in South Wales in the 1850s. The more directly coupled long crosshead type followed, many hundreds serving iron works everywhere, together with tandem layouts both vertical and horizontal. The quarter crank engine with the steam and air cylinders on separate cranks set at 90^0 followed, and almost every possible placement of the parts was tried out, as with the Perry's of Bilston engine.

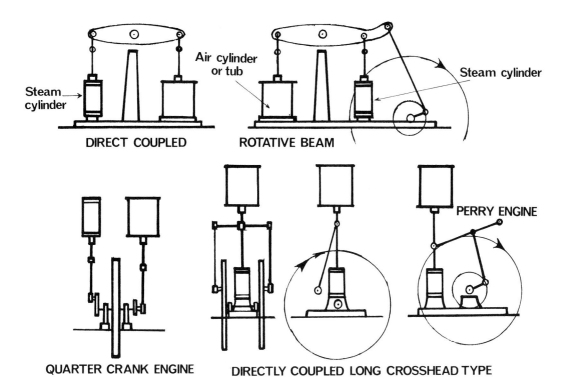

Reversing Gear

The valve within the valve chest is driven through the reversing link by either of the eccentrics, to drive the engine ahead or in reverse. The sketches show the valve gear and valve in mid-position when no steam will be admitted. Movement of the reversing link by the reversing rod and drag link will move the valve to admit steam for the required direction. In the Stephenson gear only the reversing link moved up or down to change direction. In Allan's motion the valve drive rod and the straight link move in opposite directions to reverse. In Gooch's motion only the valve drive rod is moved up or down in the reverse curved link to change direction. In Joy's gear there are no eccentrics, the motion being derived from the connecting rod by the links to the curved guide, which is tilted to the right or left to change the engine direction.

Rolling Mill Engines

With the duration, load and speed varying at each pass, rolling mill engines required manual control and the men were artists in handling the engines and placing the metal in the right roll groove. Piston valves were almost standard in later years, driven by a link motion, but Joy's valve gear with the valves on top gave a compact and almost monobloc engine. In rolling everything came second to the rapid manipulation of the workpiece, and economy, badly needed in view of the scale of operations, was difficult to secure. One method was to have the entire working of the engine controlled by a single lever and valve gear alone, leaving the steam supply valve open. With the lever in mid position and the engine stopped, a slight movement either side set the engine turning slowly in the required direction, and as the rolls nipped the metal, putting the lever far over gave full power at once. Response was equally prompt at the end of the pass, since control by the valves eliminated valve chest volume influence. This system was adopted for the great Mesta engines, with link motion and linked stop valves. It was very effective with Galloways and Markhams engines with Joy's valve gear, an essential feature being power servo-systems for rapid movement control.

Engine framing developed to meet the stresses of deep winding, and of rolling large ingots. For many years the plain section beds (Plates 16 and 19) gave good service and maintained alignment, but often failed if the foundations sank or stresses increased. Trunk beds (Plates 28 and 31) were an improvement, with less metal better placed, and sufficed for winders and continuous mills, but for heavy reversing engines beds such as Plates 80 and 81, virtually deep girders of immense strength, resisted every effort at breaking them.

Valves and Valve Gearing

It was essential that winding and reversing rolling mill engines should respond rapidly to handling requirements. In later years fuel economy was even more important, so a brief review of steam systems is desirable. The steam events within the cylinder were controlled either by a single slide valve or piston valve for winding and rolling engines, together with four Cornish, Corliss or drop valves for winders. Their action together with the direction in which the engine ran was usually effected by one of the link motions, but Joy's radial gear was also used for rolling mill engines. Slide valves and piston valves were usually driven by a single valve rod, the four Cornish valves by toes or lifters on cross shafts, with the Corliss valves and some drop valves from an oscillating wrist plate. A few drop valves were driven from a rotating side shaft.

Within the broad framework of being able to work readily in either direction, the duty of the rolling and winding engine was entirely different. The winding engine ran the same distance with a similar loading in each cycle, accelerating rapidly and stopping at the same point to land the drams. The rolling mill, since the workpiece became longer and smaller at each pass through the rolls, varied in load and revolutions at each pass.

Variable expansion gear was fitted to many winding engines to provide for the varying load. For slide or piston valves, an auxiliary drop valve or piston cut-off valve was fitted before the valve chest, and often cut-off valves on the main valves themselves, or internal cut-offs in piston valves. Cornish, Corliss and drop valves were fitted with every possible trip or releasing motion. Controlled by a speed governor, these often reduced the cut-off by the fourth revolution of the engine.

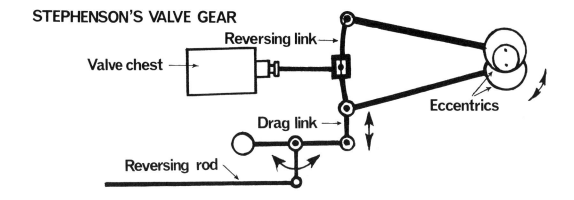

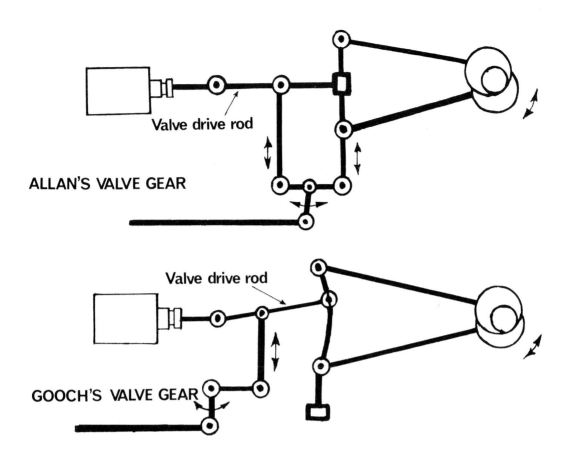

ALLAN'S VALVE GEAR

Valve drive rod

GOOCH'S VALVE GEAR

Valve drive rod

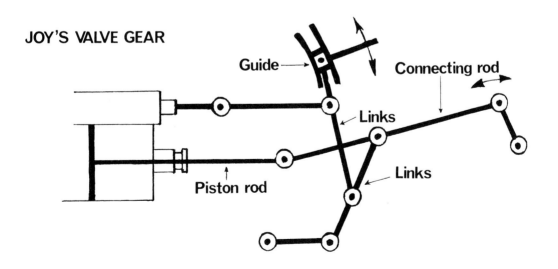

JOY'S VALVE GEAR

Guide

Connecting rod

Links

Piston rod

Links

Mine Winding Engines

Horse gins, in which the rope was wound upon a drum on a vertical shaft turned by horses, were used in every mining area. In the eighteenth century a colliery 360ft deep, with eight horses working in shifts of two each, raised 100 tons of coal daily. The advantages in speed and cost of mechanical winding led to the use of Newcomen engines to pump water to waterwheels for winding where a natural supply was short. Direct winding by steam engines developed rapidly from the use of Watt's rotative engine at Wheal Maid, Cornwall, in 1784. This replaced forty horses, winding as much in one shaft as horses did in three shafts, at much less cost, and also driving crushing stamps when not winding. Newcomen engines, driving directly, were extensively used for small pits in South Staffordshire. They were handled by boys, and would work out several shallow pits from a single engine. A Newcomen winder worked at Farme Colliery near Glasgow until 1914.

The first breakaway from the beam engine soon followed with Phineas Crowther's vertical type, patented in 1800. This simplified the whole engine by placing the rope drum overhead and driven directly from the crosshead. The last single cylinder examples were built in Leeds in 1890, and the last Durham made one ceased working in the 1960s. The pure Crowther type with a single cylinder and parallel-motion crosshead guides was usually condensing, and with cylinders up to 68in bore they were readily handled by men who knew them, to develop up to 200hp and raise 90-100 tons per hour from 1,800ft. They needed few repairs and were economical. Many vertical winders were used in other areas but they were usually non-condensing twin cylinders using steam at 60-80lb/in^2 compared with 20-30lb/in^2 in Durham.

A horizontal engine was used for winding by Richard Trevithick around 1800, and from the use of compound horizontal engines by Sims and West in Cornwall in the 1840s, single or double cylinder non-condensing engines became a standard winding type and were built into the 1930s. A small number of inverted vertical engines, ie with the cylinders overhead and the rope drum at floor level, were also built around 1880-1926 for winding.

The importance of winding meant that once a sound type evolved there were few experiments in the basic design, but there was continuous development of the details. These included advanced braking systems and controls for landing, expansion cut-off gears to save steam, new cylinders to use higher steam pressures, and steam-hydraulic servo-systems for the controls. The latter were a boon to the enginemen, as the continuous operation of valve gear and brakes was very tiring. It is an immense tribute to the men that they made untold millions of winds, doing an exacting and tiring job by hand control with so few mishaps during nearly two centuries. The automatic landing and braking systems operated very rarely.

In later years, the trunk type of frame and guides was developed, which could be readily finished with accurate machine tools thus avoiding the hand fitting of the older flat bed and four-bar guides.

Steam valve systems developed from a single flat slide valve, or piston valve per cylinder, to four Cornish drop valves at one side or in the corners, four Corliss semi-rotating valves at the top and bottom, and drop valves at the top and bottom, often with exhaust valves differing from the inlets. These were usually driven by Stephenson, Allan or Gooch link motion, but Robey and others drove the drop valves from rotating side shafts, which made the engines very light to handle.

Winding engines performed a difficult duty in that at the start of a wind the engine had to start against the load of the mineral and the rope down the shaft, accelerate rapidly, and end the wind speedily and accurately. The out-of-balance effect of the rope could be reduced by using the tapered rope drum patented by John Budge in 1772, in which the rope wound first on the smallest diameter which increased as the wind progressed. Another system was two or three parallel turns on the smallest diameter, with a rapid rise to the maximum diameter. The large single engines of the North East were designed for flat ropes which gave counterbalance as they wound turn upon turn, so increasing the coil diameter during the turn. Many were also fitted with the counterbalance system outlined in Plates 8 and 9. This was occasionally fitted to horizontal single winders, as at Usworth

Colliery (Co Durham) in the 1860s.

The great cylindro-conical drums for deep shafts were very heavy and costly, weighing up to 120 tons with shaft and cranks, and this had to be started and stopped at every wind. The Koepe system in which the rope drives only by traction, simply wrapping halfway around the driving drum needs but a single narrow pulley since no rope is wound on, but a tail rope is used, and up to four ropes side by side for safety.

The tail rope was the simplest counterbalance, comprising a rope attached to the bottom of one cage, passed down the shaft around a bottom pulley and up to the other cage. If of the same weight as the main rope, this equalised the out of balance load. Despite a few mishaps, it dispensed with up to 30 per cent of the boiler load for the winders.

Unusual conditions sometimes led to special designs for winding engines, especially in metal mining, where fuel was scarce. A single large compound condensing engine often wound from several shafts by drums connected by clutches and gearing. The engine called *Superior* at the Calumet and Hecla copper mine in Michigan, USA, was an example of this. The engine, with cylinders of 40 and 70in bore x 6ft stroke ran condensing on steam of $135lb/in^2$ and developed 2,300hp at 35rpm to 4,100hp at 60rpm. It drove a mainshaft totalling 130ft long, and 18 to 14in diameter, from which six winding drums were driven through band clutches and gearing from one side of the engine and three air compressors off the other side. It also drove by belts and then by wire ropes to a crushing mill and also to a driving unit for a man engine and a line of pump rods on a 40^0 slope. The mills and man engine unit were each several hundreds of feet away. Built about 1881 it gave many years of service.

The need for faster winding from ever greater depths in the copper mines of the USA led to the use of very powerful simple and compound double-diagonal engines. In South Africa, compound condensing engines were widely used, often with double, clutched drums, when veins of ore at different depths were worked from one shaft. The Whiting system, using a single rope travelling over twin driven pulleys to both cages, also gave good service.

A point to note is that the change from the slower condensing single cylinder engines to the fast non-condensing twin cylinder engines soon exposed the wastefulness of the new type, and engineers thought there might be a return to the old. The Durham condensing Crowther engine certainly used the least fuel per ton raised until, early in this century the mixed pressure turbine gave as much, if not more, power from the exhaust steam as did the winder itself.

My examples are of engines for large mines with ample reserves and capital. There were, however, many small mining ventures where simpler plant was needed: **a** when a small group with limited capital was trying out or re-working an area; or **b** for a difficult up-country site abroad. For **a** the cheapest was a simple 'donkey' or jigger, a vertical boiler and engine with gear-driven drum, mounted on girders. Readily stripped and re-assembled, Appleby offered a 25hp set that weighed 10 tons and cost £425 in the 1890s, and a complete mine top plant with pump for £600. For **b** the semi-portable plant developed in the 1870s by Robey was one of several suitable units. For export, Robey provided a wrought iron base comprising five strong tanks into which the engine parts, gearing, drum and spares were packed, making with the boiler six strongly built sections easily re-assembled on site. A 65hp set weighed 16½ tons in all. Such a Robey semi-portable engine was used to sink the 12½ft diameter, 1,500ft-deep Florence colliery shaft near Stoke-on-Trent. The twin cylinders were 16in x 2ft stroke, and the shaft was sunk at 25ft per week. It then wound 170 tons of coal per 8-hour shift, from 1,320ft, until the permanent winder was ready, and later drove the pit sawmill.

One often met the engine that 'came off a ship — it was a ship's engine'. This was scarcely ever correct, but it *was* so with John Nixon when he started his 'Navigation' pits in South Wales in the 1870s. The oscillating and sidelever engines did come from ships being broken up, but one, a Maudslay, of 1861, came from HMS *Gibraltar*, on her conversion in 1872. The cylinders of 82in bore x 4ft were the largest used in a winder, and it ran until 1896; tribute indeed to the men who made such odd engines do winding to exact levels at each wind. In further tribute to the winding enginemen, during the period 1887 to 1891, following the Mines Safety Act of 1887, some 937 million tons were raised in mineral mines in the 5 years, with only twenty-one lives lost by overwinding. So, almost entirely on hand control they made some 450 million winds for mineral and many millions more with the men, with a dozen or so fatalities.

Beam Engines

The basic outlines of the beam winding engine was little changed during the century in which it was constructed. My examples show some variations of the details, but the *Frontispiece* and Plate 3 fairly resemble the first one of 1784 by James Watt for Wheal Maid in Cornwall.

Frontis Prestage and Broseley Tileries, Deep Pit, Broseley, Shropshire　　　　　　　　　　1937
Broseley Foundry, c1830? Raised 12cwt of fireclay per wind 300ft. 15in x 3ft. Slide valve, condensing. 15-20hp? 25rpm. 12lb/in^2.
This was the simplest possible engine, with a timber 'A' frame to support the beam, within a simple wooden house. It was a double-acting engine with a slide valve operated by buffers on the air-pump plug rod. There was no reversing gear, the engine being started, and often run through the wind by hand levers on the weigh shaft. The simple rectangular connecting rod was offset backwards towards the crank pin, on the lines of the Heslop engine, with the beam, square crankshaft, flywheel and circular ribbed drum-shaft, and gearing all of cast iron. Fitting to the shafts was by large clearances filled with hard wood packing and steel wedges. An egg-ended boiler 4ft 6in diameter and 20ft long fired by waste coal supplied steam and it was fired by the engine driver.

1　The Lilleshall Co, Priors Lee Pits, Shropshire.　　　　　　　　　　1937
St George's Ironworks?, c1830. Duty unknown. About 36in x 6ft. Cornish valves, condensing. 10-15lb/in^2.
The date 1830 was cast on the beam, but no maker. The valves were coupled to open fully for the whole stroke, and there were primitive cone and bridle fittings where the rods entered the valve chests. The valves were operated by buffers on the air pump plug rod, and hand reversed. The beam was 18ft long between end centres, the flywheel 14ft diameter, the gears 4ft and 7ft 6in diameter, and the rope drum about 10ft diameter. As far as possible, the entire engine was of cast iron.

2　Sevens Pit, near Walsall, Staffordshire. Pit closed 1952-3?　　　　　　　　　　1955
Maker, date, and duty unknown. About 24in x 5ft. Cornish valves, condensing. 10-15lb/in^2.
Although of local make this engine was attractively light and neat, suggesting early date — possibly 1830s. It had been worked hard and suffered from subsidence, needing cross braces between the columns. Extra weight had been added to the flywheel rim. The Cornish valves were placed one above the other in a single chest at the top and bottom, again operated by plug-rod buffers, and hand reversed. The spring beams and cross entablature were of timber, otherwise, except for pins, it was almost entirely of cast iron. The beam was 15ft long between end centres, the flywheel 14ft diameter, and the 6ft rope drum was driven by 3ft 6in diameter equal gears. It almost certainly had reels and flat ropes originally on the square cast-iron crankshaft.

3　The Oswald Pit, Craghead, Lanchester, Co Durham.　　　　　　　　　　1953
Dunston Engine Co, Dunston-on-Tyne, 1897. Raised 2 tons 470ft. 32in x 6ft. Piston valve, non-condensing. 60lb/in^2.
This was the survivor of the last four beam-engine winders, and gave over 60 years hard service. The beam, entablature, spring beams and connecting rod were of steel, but diagonal struts had been inserted to the inner vertical column. The flywheel and cast iron drum were made in halves, with the 12ft 6in drum directly on the crankshaft. The piston valve was driven by a lattice rod from the slip eccentric, being reversed at each wind by hand operation with an unusual jointed handle to disengage the valve. It was in every way a fitting last example of the type and was very fast and effective.

Vertical Single-Cylinder Winding Engines

Patented by Phineas Crowther in 1800 this was the first breakaway from the beam engine. The crankshaft and rope drum were placed overhead, and driven directly by a crosshead and connecting rod, with the inner bearing supported by a massive timber 'A' frame at first, and later by a cast or wrought iron framework or masonry. The crosshead was guided by a three-linked parallel motion, often pumping by an extended motion beam as in Plate 4. The Cornish-type valves were driven from a plug rod, and often worked entirely by hand. Most widely used in the North East, sheer expertise ensured that they did not stop on centre. Later fitted with Mines Act safety equipment, they served for 1½ centuries, the last, now preserved at Beamish, stopping in 1963.

4   Fortune Pit, No 1 shaft, Burnhope Colliery, Lanchester, Co Durham.   1948
Engine House and and Headgear.
This was almost unchanged since it started in 1846, except for the round ropes, and pithead rearranged for later screens. It was a tribute to the men who designed and built it so soundly. The pumping beam can be seen projecting through the wall, and the exhaust pipe at the side. The timber headgear was a traditional small pit type, and the arched window at the top provided for rapid crankshaft renewal, and was usually bricked up. It was impressive even in the dull wintry day of my visit.

5   Fortune Pit, No 1 Shaft, Burnhope Colliery, Lanchester, Co Durham.
Thomas Murray and Co, Chester-le-Street, 1846. 20 tons per hour, 450ft 27in x 5ft. Cornish valves, non-condensing, 50hp? 40lb/in^2.
This was a typical Murray small engine, with rectangular valve chests, open type valve- operating shaft brackets, and cone and bridle valve spindle packings. All was unchanged, despite the long hard work it did. The circular buffers that worked the valve levers are seen on the plug rod, and above is the massive indoor jaw end of the pumping beam, (the last remaining example of this feature). The attractive gussets for the cylinder flange, the sharp angle of the side pipe bend, the smoothness of the castings and fine forging testify to the high standards of Murray's craftsmen. The sectors on the valve arbor shafts are seen locked by a sliding tongue to prevent engine movement.

6   Fortune Pit, engine middle level.
Murray's connecting rods were fine forgings with a neat swell and centre boss. The simple parallel motion, with the plug rod driven off the light upper beam, and massive lower pumping beam are seen as well as the very simple swinging links of the motion. The round rope drums almost certainly replaced earlier reels and flat ropes. The simple two-piece flywheel was 20ft in diameter. The inner crank shaft bearing was supported by the stout inclined timbers of 15in square pine, joined by intricate scarfed joints. These timbers (seen beside the flywheel), were joined at the top to a similar cross timber and all held together in a massive casting seen at the top. The brake track was between the drum halves.

5

7 South Hetton Colliery, South Hetton, Co Durham.   1948
Thomas Richardson, West Hartlepool, 1851. 45in x 6ft. Cornish valves, condensing. 300hp. 30lb/in².
This shows the general similarity of the Durham vertical engines, with jaw-ended cast-iron motion beams (not for pumping at South Hetton), bossed connecting rod and simple swing links. This one was powerful and the framing was of double 12in x 12in square pine timbers, and the plate shows the top junction casting. The brake acted on the flywheel rim.

8 Wearmouth Colliery, Sunderland, 'A' Pit.
Thomas Murray and Co, 1848. 61in. x 7ft. 1800ft deep.
Plate 8 shows the superb engine house and to the right, the shaft headgear and backstays. The counterbalance system to the left comprised a small pit (the staple) half of the depth of the winding shaft, in which the twin flat-ropes, each with 5 tons of chain at the end, (seen at ground level), rose and fell. The ropes were attached to twin reels on the crankshaft, and when starting the wind this was wound off the reels so as to oppose the weight of the rope down the pit. All of this was wound off down the staple pit by mid-wind, when the winding rope weights were equal, and the balance ropes and chains were then wound on again in oppostion to the shaft ropes, offsetting their increasing out-of-balance weight. By the end of the wind, the whole was again ready to assist the engine.

9 Wearmouth Colliery 'A' Pit. Rope drums.
The main ropes wound coil on coil, between the 'horns' on the drum, away from the camera, while the counterbalance ropes on the small drums came back overhead to the staple pit headgear.

Vertical Twin-Cylinder Winding Engines

Twin cylinders driving cranks at 90° on an overhead crankshaft gave a fast responsive engine, and were used in most coalfields. The framing was usually of cast iron or masonry (frequently cast iron 'A' frames) supporting the crankshaft, with parallel motion, four bar, or bored crosshead guides. Usually plain but sound, every type of valves was used, almost always worked by link motion. Counterbalance was not used with this type, but conical drums were fitted occasionally.

10   Rockingham Colliery, South Yorkshire.
Believed Lilleshall Co, 1895? Designed for 600yd wind, later 200yd. 40in x 6ft. Cornish valves, non-condensing. 100lb/in^2.
An attractive engine in a fine brick house, the framing comprised massive diagonals on each side joined by a neat cast iron strut. Like the timber 'A' frames, the diagonals were carried down to cylinder level and the walls. The three-link parallel motion was very simple, with plain plate beams and no valve gear or pumps to operate. Allan's straight link reversing gear drove the valves by rocker shafts and curved wiper toes. The delightful acorns turned on to the tops of the valves operating rods in the foreground added charm to an engine that was a credit to its builders and operators for over 60 years.

11   Sandholes Colliery, Walkden, Lancashire, No 1 and 2 Shafts.   1960
Nasmyth Wilson and Co, Patricroft, 1866. Duty unknown. 30in x 5ft. Slide valves.
This was a rare layout, with two vertical winders, and a vertical capstan engine in one house. Ashlar walls supported the crankshafts and drums above. The plain sound design calls for little remark — all was simple and functional, although the angle section crosshead guides did contrast with the usual 'T' section. The driving position on the top floor allowed both the enginemen to see the cages come to bank.

12   Sandholes Colliery, Engine House for No 1 and 2 Shafts.
In contrast to the engines, great care was taken to make the engine house attractive. The tall windows had multiple panes of glass set in frames that were artistry in cast iron, and gave a very light house. The large wall area was broken by two reeded bands of brick, with attractive dentils in the round tops of the windows. Matching dentils in the top band of brickwork, and stepped corbels supporting the roof all round, together with a delightful louvred roof ventilator, reminded one of the chapel-like engine houses of South Wales. The exhaust steam passing through the circular feed water heater outside escaped in quiet puffs that belied the busyness of the engines.

13   Wombwell Main Colliery, South Yorkshire.   1951
J. Musgrave and Son, Bolton, 1853. 30in x 5ft. Cornish valves, non-condensing.
A Lancashire engine working in a Yorkshire pit, this was an early example with brick walls, 36in thick, supporting the crankshaft. The valves were operated by Gooch link motion and later trip releasing gear was fitted to the inlet valves. It was probable that new cylinders or higher steam pressure was adopted during its century of activity, originally using 25-30lb/in^2 of steam. The castings supporting the guide bars were built into the house walls. The engineman was typical of the sturdy, conscientious men who spent their lives making a staggering record for reliability. With millions of winds per year, often 500 in a single morning drawing shift, failures were few indeed.

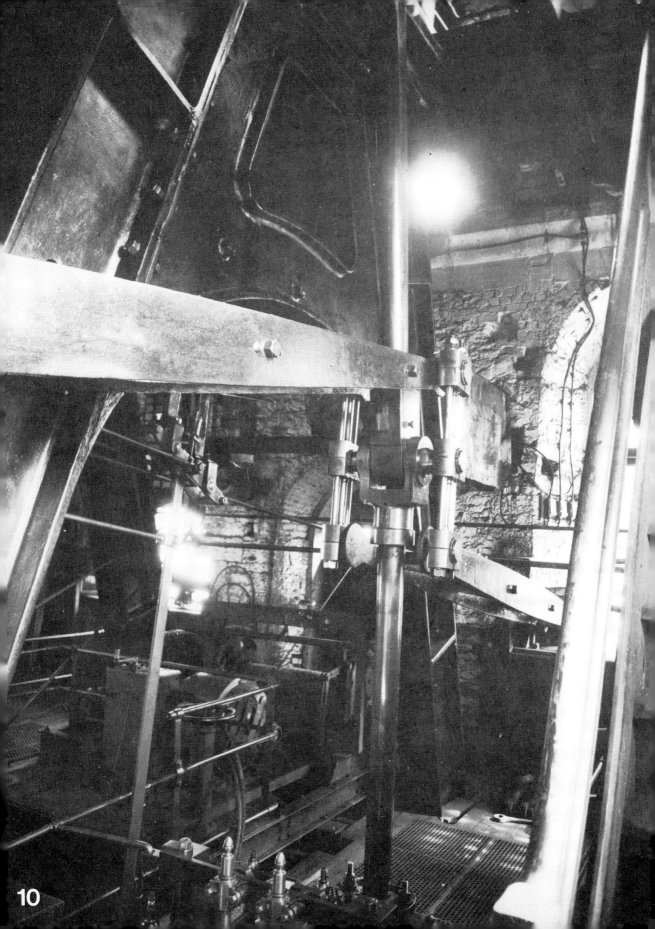

Inverted Vertical Twin-Cylinder Winding Engines

This was a later development, with two cylinders placed overhead, not widely used in mining. The largest simple expansion winder built was of this type. Made by Fowler of Leeds for Harris Navigation Colliery, South Wales, the cylinders were 54in bore x 7ft stroke, and it ran for 40 years. Although this type was covered in Phineas Crowther's patent of 1800, the earliest record I have of its use in winding is at Crawshay's Castle Pit of the 1860s. This was a 36in x 4ft 6in twin cylinder engine with Cornish valves, flat ropes running on reels, and raised about 3 tons from 1,000ft deep. The steam pressure was 45lb/in^2. The framing comprised four castings to each side, with parallel motion to guide the crossheads. Castle pit had a steam assisted brake. The latest I knew was a Bradley and Craven engine of the 1920s for Wath Main colliery, a 38in x 5ft twin cylinder engine with drop inlet and Corliss exhaust valves. This used steam at 120lb/in^2 to raise 4 tons of coal from 1,800ft deep in 45 seconds. It had bored guides, was 27ft high with cylinders at 21ft centres, and a 17ft diameter parallel drum. All records and drawings have long disappeared.

14   Tirpentwys Colliery, No 1 Shaft, Pontypool.   1960
Daglish and Co, St Helens, 1905. Duty unknown. 40in x 6ft. Drop and Corliss valves, non-condensing. 120lb/in^2.
This impressive engine was 25ft high in a house limited by high ground behind and the nearby shaft. Bored crosshead guides in the stiff mid section were joined by a flange to the base by widely spread column feet. Drop inlet and Corliss exhaust valves were fitted in the cylinder heads, with governor controlled cut-off gear, driven by Gooch link motion. The steam reversing engine was on the middle platform near to the right-hand engine. The cylinder lagging was very neat, and the cylinders were fitted with cross ties.

15   Elliott Colliery, near New Tredegar, Glamorgan, West Pit.   1965
A. Barclay and Co, Kilmarnock, c1896. Shaft 550yd deep. 34 and 56in x 4ft 6in. Corliss valves, exhaust to turbines.
Built as a cross-compound engine this, too, had bored crosshead guides but with the frame comprising a single casting for the front and back, with forged cross ties. Allan link motion controlled the valves which were placed in the cylinder barrel. It was fitted with a cylindro-conical drum, later changed to parallel with a tail rope, but the result was not very good.

Horizontal Single-Cylinder Winding Engines

This was the simplest of all winders, widely used in small collieries. My examples are from Durham, where single cranks were widely used, and a Lancashire pit where sheer shortage of capital dictated the type.

16   Burnhope Colliery, No 2 shaft, Lanchester, Co Durham.   1949
J. Joicey and Co, Newcastle-on-Tyne, 1868. 36in x 5ft 3in. Cornish valves, non-condensing. 35-50lb/in².
A typical sturdy Durham engine, this was the main drawing unit in later years. It remained little altered for some 80 years, except for metallic rod packings and expansion gear added later, and Mines Act safety gear. The 17ft flywheel comprised two half castings with four arms each, joined by hoops on the hubs, and eight rim sections fitted to the arms by dove-tails packed with timber and wedges, and bolted rim joints. The landing marks on the rim show how these were arranged to stop with the crank at mid-stroke by altering the drum lagging. There were three coloured bands, but no boss on the strap-ended connecting rod, and the castings were all small and easily handled. It was a credit to the men who built and ran it.

17 Adam Mason and Son, Montcliff Colliery, No 1 shaft, near Horwich, Lancashire   1966
Maker unknown, c1872. 390ft deep. 15in x 3ft. Slide valve, non-condensing. 50lb/in².
Set upon a hillside, this little pit, with the main winder a 14in x 1ft 6in twin cylinder engine winding one tub of coal per wind, was a good example of many years of struggle to keep a small private mine in being. No 1 was a small factory-type engine driving the 6ft curved-spoke rope drum by 1½ to 1 gearing with Stephenson's reversing gear. The single cylinder Davey differential pumping engine is seen at the other side of the engine house. Installed in 1881 the cylinder was 24in x 4ft non-condensing, and it drove a plunger pump for 60 gallons per stroke by 10in x 10in timber rods to the pit bottom. This had cost £410 in all, a severe blow to the little enterprise which nearly went bankrupt from the cost, but it struggled for nearly a century, latterly as a licensed pit. The headgears were of timber, and little had been altered.

Horizontal Twin-Cylinder, Slide Valve, Winding Engines

These were the simplest and handiest engines, readily made in the small engineering shops that were such a feature of the Victorian era. They were usually built upon a single bed plate each side, with four-bar crosshead guides, forged strap-ended connecting rods, and simple link-motion reversing gear. Control was by steam throttle and reversing levers, with a simple foot brake for the older ones. They thus tended to look alike, but my examples indicate that most makers had individual features.

18   Old Mills Colliery, The Old Pit, Radstock, Somerset.   1966
William Evans, Paulton Foundry, Somerset, 1861. 25in x 5ft. Slide valves, non-condensing.
Paulton Foundry was a typical small country shop serving the pits with work outside their capacity. The valves were in half-round chests on top of the cylinders, with cast-iron drum ends, square crankshaft and the driver on a high platform able to see the cages land over the drum. The link motion was most unusual, since the eccentric rods were supported on swinging links, with extension rods to the link. This meant much less weight to move in reversing, ie only the link and two short rods. The steam brake was an addition, but it was otherwise little altered.

19   Morton Colliery, Clay Cross Pit No 5, Derbyshire.   1964
A. Handyside and Co, Derby, works numbers 88 and 89, 1865. 30in x 5ft. Slide valves, non-condensing. Raised 3½ tons of coal 1,750ft.
Two interesting features were: the Allan link motion was driven by a link pin in the connecting rod, which reduced the width between the crankshaft bearings and made the eccentrics more accessible; and the use of hollow box-type valves sliding between machined surfaces, which reduced the load on the valve gear. The piston tail rods had been removed, an early Clay Cross Company practice. Other features were the unusual rectangular connecting rod section, the cast-iron drum ends and cast-iron roof trusses carried down to brickwork below the short windows. Little was altered except Mines Act fitments. This wound up to 1,200 tons of coal per shift.

20   Cwm Colliery, No 1 and 2 shafts, The Ebbw Vale Co, Glamorgan.   1962
Nasmyth Wilson and Co, Patricroft, works numbers 587-8, 1891. No 588: 41in; No 587: 36in x 6ft. Four slide valves per cylinder, non-condensing.
Virtually unaltered in 70 years heavy work except for some metallic rod packings, these were remarkable for the separate inlet and exhaust slide valves in separate chests. The valves were backed by set screws on the back. The engines were 40ft long over the bed plates, and one was seriously damaged when a fire occurred in the brake lining in 1953, and firemen sprayed water on the heated bed. This resulted in multiple cracks in the beds and Metalock stitching was effected, at first to allow reduced winding, and finally, over weekends, the whole was stitched and the engine ran for two more years until electric winding was installed about 1956.

21   Nailstone Colliery, near Bagworth, Leicestershire.   1970
Worsley Mesnes Co, Wigan, 1921? 24in x 4ft. Piston valves, non-condensing. 510ft deep.
An attractive engine by an uncertain maker, photographed a few weeks before closure and scrapping. The inside admission valves were driven by Stephenson's link motion, and were in separate chests bolted on. No features suggested the makers, but it could have been de Ath and Elwood who supplied winders to the nearby Ellistown colliery. The Worsley Mesnes Co certainly did work on it (possibly new cylinders?).

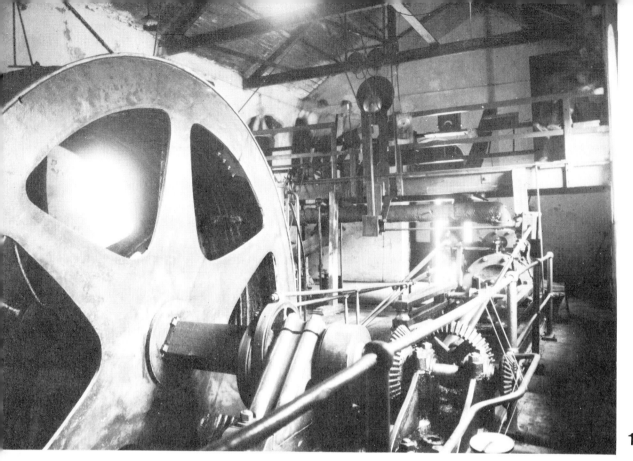

18

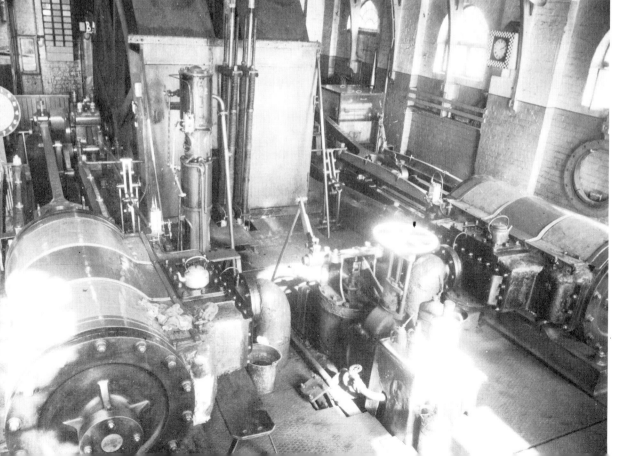

19

Horizontal Twin-Cylinder, Cornish and Drop Valve Engines

These two types of valves are similar, ie circular and of the equilibrium type. The Cornish valves, working in pockets or chests bolted to the cylinder barrel, were either in pairs at one side, or at the corners, as Plate 23, operated by toes or wipers on swinging shafts. Drop valves were placed at the top and bottom of the cylinder, and often driven from eccentrics on rotating side shafts or lay shafts. They were easier than slide valves for manual handling, but however well adjusted they were usually noisy.

22   Abercynon Colliery, Glamorgan.   1953
John Fowler and Co, Leeds 1891, works number 6209. 42in x 7ft. Cornish valves, non-condensing. 100lb/in².
This had new cylinders and cut-off valve gear fitted by Daglish, St Helens, possibly about 1912-14. Named *Fair Rosamund,* with four-bar piston and tail-rod guides, it had new cranks and crankshaft fitted by Markham in 1956. One rear cylinder cover with Fowler's name cast on was fitted to the new cylinders. Allan's link motion drove the valves.

23   Wood Pit, No 4 shaft, Tyldesley, Lancashire.   1966
J. Stevenson and Co, Preston, c1890. 500 yards deep. Originally 36in x 6ft. Cornish valves, non-condensing. 90lb/in².
Stevenson only built two winding engines (among numerous other types) and the other one was made for the nearby Boston Pit. The Wood Pit engine was said to have been made for a pit near Southport, and moved to Wood Pit in 1934. Wood No 4 was linered down to 30in bore for steam at 145lb/in² in 1940. It raised a payload of 4½ tons of coal from 500 yards in 35 seconds, with 25½ revolutions of the drum, with steam shut-off at ten revolutions.

24   Chanters Colliery, No 2 pit, Tyldesley, Lancashire.   1965
Greenhalgh and Co, Atherton, 1896. Raised 3 tons from 600 yards. 28in x 5ft. Cornish valves, non-condensing. 100lb/in².
This was the only engine I met with Cornish valves in one chest across the cylinder centre line. Governor-controlled cut-off was fitted, with Gooch link-motion reversing gear. The rope drum was 18ft in diameter, with cast-iron spiders in two halves, and some repairs to the arms. Always well kept, it had been well cared for. The inlet and exhaust valves were driven by toes or wipers on the same cross shaft.

25   Woodhorn Colliery, Ashington, Northumberland.   1971
Grant and Ritchie, Kilmarnock, 1900. 800ft deep. 28in x 6ft. Cornish valves. 100lb/in².
There were fourteen boilers when all the colliery's winders, compressors, generators and fan were steam driven. The two winders were similar, without cut-off motion, and one may have had new cylinders in 1950. Gooch link motion. A unique feature was that the depth indicator dial rotated, with the depth shown by a fixed hand at the top. Well made, kept and maintained, the original cast-iron drum structure was intact after 70 years of hard work.

26   Wearmouth Colliery, 'C' shaft, Sunderland.
Bever, Dorling and Co, Bradford, 1890s? 1,500ft deep. 34in x 5ft 9in. Cornish valves, Allan's link motion. 650hp. 30rpm. 120lb/in².
This was the massive four-bar crosshead guide design which Bevers adopted for deep shafts prior to the trunk guide frame. The engine had given many years of very hard service with little repair, and still retained the soft gland packing for the piston rods but it may have had an automatic steam cut-off system fitted later. The rope drum was 24ft in diameter, but without the counterbalance system fitted to the 'A' and 'B' engines. The exhaust steam was utilized by a turbine compressor of similar power. Photography was difficult as the wartime blackout was still in position.

22

23

24

25

26

27

27 . South Celynen Colliery, No 2 shaft, South Wales. 1967
Robey and Co, Lincoln, Works numbers 43040 and 43041, 1926. 380 yards deep. About 24in x 4ft. Drop valves. 140lb/in².
Robey's engines were clean and neat, and well liked by enginemen for their speed and easy handling. The governor controlled cut-off made them economical. The design was that of their mill engines and gained Robeys a very wide reputation for long service and the readiness with which the few repairs were carried out. This one was very fast, with a conico-cylindrical drum.

28 Bridgewater Trustees, Brackley Colliery, No 1 Pit, near Bolton. 1958
J. Musgrave and Co, 1879. 310 yards deep. 26in x 5ft. Cornish valves. 100lb/in².
This shows the traditional Cornish valve layout at each corner of the cylinder, in separate chests bolted on. The very simple drive was by a single flat rod from the Allan link motion to the single cross wiper-shaft to operate the valves by the wipers or toes and long lifting shafts. The steam and exhaust valve drives were identical with no trip cut-off mechanism. Metallic packing had been fitted to the piston and valve rods, but other than a new drum and safety gear, little was changed.

29 Nook Colliery, Astley, Lancashire. 1958
J. Musgrave and Co, 1911. 40in x 6ft 6in. Drop and Corliss valves. 950 yards deep.
Musgrave's improved design illustrates driving drop valves and Corliss valves from a wrist plate and Allan link motion. The trip cut-off was under governor control. Piston tail rods were fitted running in tubular cases, as seen on the left-hand cylinder.

Other Types of Winding Engine

A few notes upon less familiar types to end the winding engines. The twin tandem engine comprising a high-pressure and a low-pressure cylinder in tandem for each crank, was adopted around the turn of the century where speed, high power and fast handling were required. Fitted to some of the large collieries they were magnificent engines. The cross-compound engine with one high-pressure and one low-pressure cylinder, each coupled to its own crank, was also very efficient and effective. I end with examples of very rare angle-compound engines (or inverted Manhattans), one built as such and one converted to the type.

30   Penallta Colliery, South Wales.   1956
Fraser and Chalmers, Erith, 1906. 2,250ft deep. 32 and 56in x 5ft each side. Drop and Corliss valves. 2,500hp. 140lb/in^2.
Penallta power house was a magnificent building, with a very fine powerful twin-tandem winder at each end. Using superheated steam there were drop inlet valves for the rear high-pressure cylinders, but the rest were Corliss-type semi-rotating valves. The strong yet simple frames gave the appearance of power that the colliery required, raising some 6 tons per wind in 60-70 second cycles. The conico-cylindrical rope drums rose from 15 to 25ft in diameter with several turns on the maximum diameter, making up to 25rpm. They were almost as built after half a century of heavy loading. To the right, between this and the other winder, were the air compressors, turbines and electrical generating sets, including LP turbines utilising the winder exhaust. Many Fraser winders had cast-steel disc cranks, but the largest had these forged steel ones.

31   Astley Green Colliery, No 1 Shaft, near Manchester.   1955
Yates and Thom, Blackburn, 1912. Raised 9 tons per wind from 2,500ft. 35 and 60in x 6ft each side.

Corliss valves. 160lb/in^2.
This was built, together with an identical pair for Askern Colliery, when Yates and Thom were at their peak for engine sizes and output in winders and mill engines. The valves were driven by Allan link motion and wrist plates, and the exhaust steam developed over 2,000hp in the low-pressure turbines at full load. It was nearly 70ft long overall. There were fifteen boilers at the colliery with everything on steam.

32   Elliott Colliery, No 1 Shaft, New Tredegar, Glamorgan.   1966
Thornewill and Warham, 1891. Works number 603. About 450 yards deep. 28 and 42in x 6ft each side. Corliss and Cornish valves.
This was built as a Cornish valve 42in twin-cylinder engine and when in 1895 water broke in almost as soon as coal winning began, the engine ran day and night for weeks, raising 6,000 tons of water daily in tanks. The original drum was parallel, 24ft in diameter, but later changed to a 15 to 26ft diameter diablo type.   The management was very efficient and the colliery was soon fully equipped with electricity, coke ovens and gas engines using surplus gas. High-pressure boilers were installed for electrical supply, and in 1904 Thornewills added Corliss high-pressure cylinders using full boiler pressure. The valves were driven by Allan link motion.

33   Sutton Manor Colliery, No 2 Shaft, near St Helens   1938
Yates and Thom, 1914. 1,650ft deep. 33 and 55in x 5ft. Corliss valves. 150lb/in^2. 46rpm.
The cross-compound winder used about 30 per cent less steam than the twin-cylinder type, and if arranged so that the wind ended with the high-pressure crank at mid-stroke, it was just as handy. The LP receiver (between the cylinders) was usually fitted with an auxilliary steam supply to maintain about 45lb/in^2 at all times, and occasionally a linked throttle valve between the receiver and LP cylinder assisted in isolating the LP when stopping. Excellent examples by Yates and Thom, and Fraser and Chalmers gave good service in Kent, Sherwood (Notts), Bank Hall (Burnley), Sutton Manor and Hatfield (Doncaster) pits.

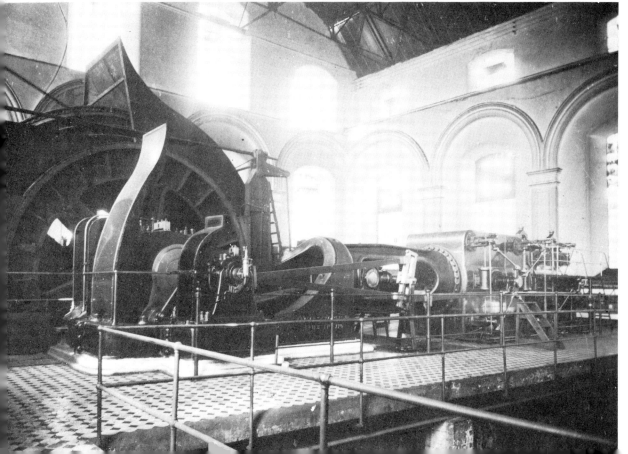

31

32

34, 35 Cinderhill Colliery, Cinderhill, Notts.   1950
The Butterley Co, 1851; Thornewill and Warham, c1890. 660-750ft deep. 40in x 5ft as built, 28in x 5ft added 1890s. Cornish valves. 40-60lb/in².
This was designed as a 200hp vertical single-cylinder engine with an iron and hemp flat rope winding upon a 13ft reel, and was aided by a 1¾ ton counterbalance winding upon a 3½ft reel. It raised 90 tons per hour and up to 3,000 tons in five days (50 hours). Thornewills added the horizontal engine driving on to the same crank in the early 1890s, and possibly replaced the original 40in cylinder by a 32in one for the higher steam pressure. It was never compound. The neat cast-iron framing was designed with diagonal forged wrought-iron stays, and the holes for the rods of the original tappet valve-gear can be seen in the top right-hand side. This was replaced by twin sets of link motion, with hand adjusted trip cut-off gear for the inlet valves at the change. It was winding from 750ft in a 75-second cycle when a century old so rapidly that each photograph comprises several exposures.

36 The Powell Duffryn Co, Bargoed Colliery, Glamorgan.   1960s
Thornewill and Warham, 1900. 1,875ft deep. Works number 834. Drum 16 to 24ft diameter. 32in horizontal and 50in vertical cylinder each side, 6ft stroke. Corliss valves. 160lb/in².
This was designed as a twin angle compound by Mr Hann the colliery company's engineer to raise 6 tons of coal in 50 seconds. The pithead had to be high to meet coal handling needs, and the angle form was adopted to give better accessibility for the pistons than did the tandem engines. The cylinders were made large to avoid the need for the supplementary steam feed to the LP receiver, said to be a source of loss in cross-compound winders. The drum rose from 16 to 24ft in diameter in three turns giving rapid acceleration, with twenty-two turns on the 24ft centre. The crossheads worked in bored guides, and each cylinder had its own Allan link motion. It was certainly a fast, powerful and handy engine.

34

35

36

Mine Pumping Engines

We really owe the development of the steam engine to the plight of the collieries when, meeting the tenfold growth of the coal trade in the seventeenth century, it was necessary to work deeper seams in the central coal basins. These rich seams were often abandoned as the quantity of water proved too great for the primitive pumps of the period. Waterwheels were extensively used in the mines of South West England, as at Wheal Friendship in Devon, where eight of their seventeen waterwheels were used for pumping. The largest of these was 51ft in diameter, ran at 5rpm, and, installed in 1844, pumped from 1,130ft deep. In a wet colliery 10 tons or more of water were pumped per ton of coal mined.

Thomas Savery's engine 'The Miner's Friend' proved at least that the power of steam could be used to raise water, but, needing steam under pressure, it was a dangerous machine with the limited constructional techniques of the times. Thomas Newcomen's engine overcame this, since it worked entirely by the vacuum produced by condensing steam within the cylinder. It proved to be wasteful, but was widely used where coal was cheap. James Watt remedied this by separating the cylinder and condenser as covered by his patent of 1769. It used only one third of the fuel consumed by the Newcomen engine. The Cornish engineers later halved this by the use of higher steam pressure, expansion and heat conservation, together with Cornish-type pitwork and pumps. Davey's horizontal engines also gave fine service in mining from 1870 to the 1950s. They were compound and triple expansion designs with the engine either placed on the surface driving the pumps by Cornish pitwork, or with the engine at the pit bottom driving the pumps directly. Usually horizontal, other underground engines were rotative or direct acting.

Pumps remote from the shaft bottom were sometimes driven by water under pressure from the main rising to the surface, and in a few instances very efficient rotative triple-expansion engines on the surface pumped water at high pressure to operate hydraulic pumps underground. The pumps were of two types: the plunger forcing the water on the down stroke, and the bucket type lifting it on the up stroke; there were many variations with features of both, such as the bucket and plunger type. The plunger type was more efficient since it was largely operated by the weight of the pump rods, rather than the direct effort of the steam pistons. This aided the Cornish engine since the pump rods were lifted by expanding steam once they started to move upwards, but the weight was constant to work the pumps on the downstroke. To reduce shock, speeds were low, rarely exceeding twelve pumping strokes per minute.

Higher speeds were achieved by underground rotative pumps abroad, by the use of very light valves such as the Gutermuth, or operating them mechanically on the Reidler system. In Britain small rotative pumps of the Evans or Pearn types also gave good service at higher speeds, in mining. So until the electrically driven centrifugal pump was developed, mines were mostly drained by slow speed pumps worked by Cornish beam, Bull or Davey engines on the surface, and Davey, Worthington and many other direct acting and rotative pumps below.

Cornwall, with mineral working rights divided into small areas, had the greatest concentration of beam engines in the world. A single engine was often arranged to wind and pump, particularly at small mines, or in sinking the shaft of a large one, often using overtype and undertype semi-portable engines. In the North East, the vertical winders were often arranged to pump and wind, by making one of the crosshead parallel-motion beams of double length, to drive pump rods in the shaft (Plate 4). The rotative Cornish beam winders or 'whims' as they were termed, sometimes worked flat rods driving to pumps in distant shafts.

Highly economical engines, driving Cornish-type pitwork through gearing, were used where fuel was costly abroad, as at de Beers' mines in South Africa. In 1889, James Simpson of London supplied them with a triple expansion Corliss engine with cylinders of 15, 23 and 37in bore x 3ft stroke which drove 70 tons of 14in x 14in pine pump rods for pumps to raise 625gpm from 1,500ft. The large gear wheel was 30ft in diameter and weighed 80 tons, with teeth 30in wide.

Beam Pumping Engines

Examples of the traditional non-rotative beam pump, varying a little from the usual in makers and usage; often long used as standby units only, little record of duties and sizes was obtainable on the sites.

37   Bagworth Colliery, Bagworth, Leicestershire.   1938
Maker unknown, believed 1820s date. 62in x 8ft. Cornish valves, pump data unknown.
Nothing was known about this engine but National Coal Board research suggested 1820. The bearing brasses and parallel motion links had Gothic ends and another early feature was putting the equilibrium valve near to the exhaust valve at the bottom of the cylinder. The valve gear had been altered to two instead of three arbors with a wing-type throttle valve in the equilibrium pipe. Another old feature was that there were three wrought-iron pump rods down the shaft. Disused for years, it was a useful standby still in the 1930s. It had been heavily used over its lifetime, having a new cylinder in 1925 when the old one was said to be worn very thin.

38   Hodbarrow Iron Mines, Cumberland.   1953
Perran Foundry, 1878. Plunger pump 24in to 300ft, bucket below. 70in x 10ft and 9ft. Cornish valves. This was probably the last Cornish pump built by Perran Foundry and was moved from its original site in 1908. The balance beam and box were added by Holman of Camborne in 1910. It was one of three engines on the site, but a shaft collapse damaged one. The beam, nearly 32ft long was 6ft deep in the middle.

39   Mill Close Lead Mines, Darley Dale, Derbyshire. Two engines.   1936
Thornewill and Warham. No 154. 42in x 8ft, 1857. No 170: 48in x 9ft?, 1860.
Two Cornish-type engines in one house were rare in mining, but Mill Close Mine was notably wet and these engines were fully used until electric pumps were installed in the 1930s. Although by the same makers, only three years apart, there were many small design differences in the parallel-motion linkage and valve gear. The strokes noted are approximate, with different beam lengths indoors. The bob plat between the beams allowed attention to either or both beams at work (as they often were). The extra straps for the pump rods indicate the heavy loading.

40   The Griff Colliery Co, near Nuneaton, Warwickshire.   1935
Two engines: Goscote Foundry *(Caroline)*, 60in x 7ft; other maker unknown, 1831, 55in x 7ft.
Two Cornish engines so close was very rare in collieries. *Caroline* had 'Goscote Foundry' cast on the beam but no date, the other had the date 1831 but no maker. Both were fitted with air-plunger cataracts above floor level, which held the engines 'indoors' (ie piston at bottom) for the pause. Both had twin equilibrium trunks, and Gothic ends to some of the motion links and brasses. The very tall rod capstan frames allowed a full rod length lift above the shaft head, with a small one for the shaftman. They were useful standby units in the 1930s.

40

41, 42   Newton Chambers, Newbegin Shaft, Chapel Town, Sheffield.   1933
Newton Chambers Co, 1856? Duty unknown. About 36in x 8ft. Cornish valves, non-condensing. Underbeam Cornish engines were rare, and this was probably built by the company for the shaft. It worked on the Cornish cycle feeding steam above the piston through the central trunk, followed by the equilibrium, to exhaust from under the piston. The use of eccentrics to operate the valves was unusual, and there was no evidence of an air pump. The arbor and wiper under the floor in Plate 42 operated the cataract. It ceased working about 1926.

Beam Engines in Cornwall

These engines were so important in the development of the county that they need a section of their own, however short this may be. The mineral deposits were so valuable that even a small 'sett' (as the leases to mine an area were termed) could support a successful venture. Each was usually an independent effort, complete with machinery for pumping, raising, crushing and concentrating its output. Water power was used wherever possible, and many horse-driven whims were raising ore throughout the nineteenth century, but the greater part was won by the aid of steam power. Native fuel was almost unknown in non-ferrous metal areas, so the highest fuel economy was needed, which led to the use of unorthodox engine types in non-ferrous mines, but in Cornwall the beam engines played the greatest part. Certainly the pumping engine reached its highest development in Cornwall, where the pitwork (the system of rods that operated pumps as deep as 2,000ft) reached incredible proportions, and often of great complexity when the shafts followed the twists of the veins in the native rock.

  The skills and expertise of the Cornish engineers modified the basic Watt cycle especially for the reduction of fuel consumption, and they reached the peak of heat conservation by steam jacketting the cylinder, using thick insulation, and various arrangements of the steam ports and valves to reduce heat loss during the cycle. The sheer density of the mineral veins over large areas of Cornwall led to many mines operating closely together, leading to a proliferation of engines unequalled anywhere else, a dozen or more engines often being visible from one point. When, in less prosperous times, smaller ventures failed and were brought under a single control as 'Consols', great economy was often secured when a single large engine replaced several small ones. This allowed profitable working of less valuable mines by lower costs, reducing capital outflow for fuel, and gave employment for the engine makers for new engines, or the removal of sound existing ones. The engines were of two main types. The draught or pumping engines gave a simple reciprocating motion from the beam to the pump rods; the rotative engine either ran in one direction to drive the ore crushing stamps, or in the whim, reversed at each wind to raise ore from the mine bottom. A few rotative engines also drove pumps through gearing.

43 Engine houses at Wheal Peevor near Redruth.
This is a typical set of Cornish mine engine houses. Wheel Peevor, once a part of the North Downs mines, was working down to 135 fathoms in 1865 (a fathom is 6ft) and still sinking to lower veins. The centre house contained the 60in x 9ft pumping engine purchased from North Downs in 1872 for £540, a fraction of its cost, and which raised the water to a drainage adit 47 fathoms from the surface. The whim or winding engine was in the house to the right, probably a 24in rotative engine raising from the full depth. To the left was the crushing stamps, a set with forty-eight heads purchased from Basset and Grylls Mine (Wendron) in 1876, powered by a 32in rotative engine. It cost £790, carriage was another £60, and the house a further £223. The pump and whim engines aligned with the shaft, but the stamps, having no direct connection with metal winning were placed at the point most suitable for ore and water supply. True to Cornish practice, each engine was quite separate with its own boiler plant and chimney.

44 East Pool and Agar mine near Camborne. Taylor's shaft 90in.
This superb pumping engine, made by Harvey of Hayle in 1892 is exceptionally heavy, weighing over 680 tons with the pitwork. When built for the Highburrow shaft of Carn Brae mine it was fitted with 8in-diameter iron rods for the first 113 fathoms, with timber rods for the remaining slant or underlay part. Moved to the new shaft at East Pool in 1924, it was then fitted with all wooden rods varying from 20 down to 16in square, to the full 1,700ft depth. John Cornwall's fine photograph, taken before additions by the now busy mine obscured it, shows one of the several balance bobs, placed throughout the length to offset the excess weight of the rods over the water loading. It shows the long metal plates fitted to spread the load on the timbers, but the swinging link coupling the balance beam to the pump rod has been removed. The piston stroke was 10ft and the pumps 9ft stroke. It continued at work until 1954.

45 East Pool mine, Michell's whim. 30in x 9ft stroke.
Designed by F.W. Micheil this was built for the site by Holman of Camborne in 1887 and is interesting for the long crank and cylinder stroke. The whim engine alternately raising and lowering the cages, has to reverse at every wind, and the engine is fitted with Gooch's link motion to drive the valves by a rocking shaft. It makes reversing very easy, but gives full stroke steam admission, and a drop cut-off valve driven by a rotating side shaft gives expansion. The rope drums are about 8ft in diameter fitted on the crankshaft; running at 17rpm each 1,500ft wind took about 3½ minutes.

46 Parkandillick clay works, near St Austell
This engine illustrates another facet of Cornish mining, the readiness to use whatever would best do the job, even to moving these great engines to meet needs elsewhere. It is a 50in engine with 10ft stroke, and was built by Sandys Vivian at the Copperhouse Foundry, Hayle, in 1852 for the Old Sump shaft of Wheal Kitty mine, St Agnes. It was moved to Parkandillick 60 years later, and, fitted with a new cylinder by Bartles of Carn Brae Foundry, pumped clay slurry until 1955. A fine example of mid-nineteenth century engineering it is a typical Cornish draught engine, and the house a good late example of Cornish masons' craft.

43

44

45

47 Parkandillick, the middle floor.
Most Cornish pumping engines had the top nozzle-chest valves arranged with the governor or throttle valve (under the driver's control) placed at one end, followed by the inlet valve (under control of the valve gear) next, with the equilibrium valve at the other side delivering to a side trunk and the lower port. Sandys Vivian however, put the equilibrium valve in the middle delivering to a central trunk, and the inlet valve at the side. It was not a good arrangement thermally, since the fresh steam had to pass around the cool equilibrium chest to the inlet. The plate shows this, the driver's governor valve at the left, followed by the equilibrium and inlet valves. Another feature shown is the automatic reduction in the steam supply if the engine made over-long strokes. It comprised a lever worked by a tappet on a plug rod which closed the throttle valve in by a ratchet motion and bevel wheels, if she overstroked when the driver was on the top platform.

48 South Crofty Mine, near Camborne. Robinson's 80in pumping engine.
Designed by Samuel Grose for Wheal Alfred Consols mine near Hayle, and built by Copperhouse Foundry, this cost £2,700 and was started in 1855. An attractive and economical engine it was moved to Crenver mine in 1864 and to Tregurtha Down in 1881, when a new cylinder was fitted following a rod fracture. She came to rest in 1903 when purchased by South Crofty for their new vertical shaft, to work day and night for half a century with the cylinder cover only lifted once. She raised over 310gpm from 337 fathoms in seven pump lifts. The plate shows a special feature of Grose's design, ie the placing of the steam and equilibrium valves at the opposite side of the cylinder, giving a uniflow steam flow and reducing heat losses from cooled ports. Most Cornish pumps were controlled by a governor or throttle valve operated by the driver and Grose, wishing to avoid the pressure loss, did not fit one. However, the sudden impact of the full steam pressure upon the piston was found to strain the pump rods and connections, and this was reduced by restricting the lift of the inlet valve (to the right) to give a gradual pressure build-up.

47

48

The Davey Differential Engine

In this type, the piston rod was coupled directly to the crosshead and guides, to drive the pumps by 'L' bobs from a timber sweep rod. They were built as simple, compound, or triple-expansion types. The special feature was the differential control of the valves by a motion compounded from that of the engine, opposed to a constant-speed cataract. By this, excessive engine speed over-ran the cataract and closed the steam valves. This prevented many accidents from loss of load by broken rods, pumps or pipes.

49 Wood End Pit, Upcast shaft for Bank Hall Colliery, Burnley.   1937
Hathorn Davey and Co, Leeds, No 3504, 1882. 210yd deep. 30 and 54in x 8ft. Slide valves, condensing. 40lb/in². Four 18in bucket pumps.
This was the short type with the cylinders close together, and twin LP piston rods passing beside the HP cylinder. The timber sweep rod was 12in square, and it cost £1,690 without shaft rods or pumps.

50 Clifton Colliery, Burnley.   1937
Hathorn Davey and Co, works number 4892, 1891. 34 and 58in x 8ft. Slide valves. Two 13in plunger pumps. 260yd deep.
This had a single rod from the HP to LP piston, and twin rods from the HP piston to the crosshead.

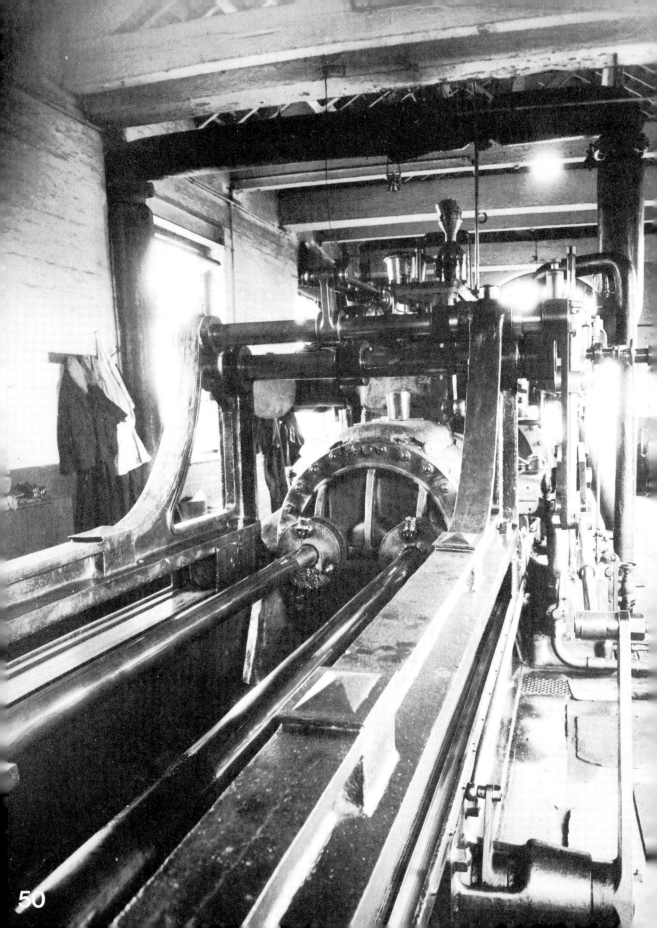

Tandem-Compound Rotative Engines

51, 52 Whitwick Collieries, Swannington, Leicestershire. The *Calcutta* Engine.   1948
Robert Stephenson and Co, Newcastle-upon-Tyne, No 328, 1877. 42 and 72in x 8ft. Slide valves. 60lb/in^2. Two 26in x 8ft bucket pumps.

This was one of the largest rotative pumps, 125ft long from flywheel tip to the end of the pump bobs. A continuous, six section, cast-iron bed went end to end, with the pump end (Plate 51) about 12ft below the engine. The 32ft 50-ton flywheel enabled it to run at the 5rpm to which it was restricted later, following fractures. It was highly finished in every way, with very neat timber cylinder casing. The five Cornish boilers by Stephensons were still insured for the original pressure at 70 years old. The engine room was over 80ft long.

Mine Ventilation Engines

Ventilation is essential in collieries, many of which would be unworkable without a strong air circulation to remove the gas discharged from the seams. For a period the air flow was maintained by a furnace placed at the bottom of the shaft, giving a chimney effect, and this was made effective for gassy pits by passing the mine air in drifts to the upcast shaft away from the furnace. The effectiveness of this system increased with the depth of the pit. The Wearmouth Colliery at Sunderland is an example of this method. It was finally sunk in 1836, the workings extended for 3½miles from the shaft by 1881, and it was over 1,700ft deep. The ventilating furnaces at the bottom of the upcast shaft had a total grate area of 144 sq ft and secured a flow of 200,000 cu ft of air per minute through the workings. This reduced the temperature in the mine workings from 90^0 to 75^0F. From the mid-nineteenth century, however, the higher efficiency of the fan, and the fact that it was of equal value in shallow or deep pits, led to the rapid development and trial of many ventilators. Certainly most of the conceivable methods of continuously moving a large volume of air at a low pressure were tried, usually with one shaft as a downcast or entry tract, with the fan or extract unit at the top of the other or upcast shaft.

The engines adopted, as so often, were a compromise. They had to be absolutely reliable, since explosive concentrations of gas soon arose from a failure of the air current. This required the simplest and most rugged engines with no complications likely to fail. On the other hand, the fact that a fan engine never ceased working involved heavy fuel consumption which required consideration particularly in later years, when fuel increased in value; fan-driving engines therefore varied considerably. A heavy single cylinder non-condensing slide-valve type running at 20 to 60rpm coupled directly to the fan shaft served many collieries, but compound engines were also used. Fans running at higher speeds were developed from about 1870 which, driven by cotton ropes or belts, could use almost any type of engine, and from 1890 enclosed forced lubrication engines, compound or triple expansion, directly coupled to fast running fans, were developed to become almost standard equipment. Although most were of the Belliss and Morcom double-acting type, single-acting engines such as the Willans, or Bumsted and Chandler designs, were also used. The performance of a Willans engine driving a Cookson fan at Woodthorpe colliery illustrates the hard work mining fans performed. Installed in 1886 this ran almost nonstop for three and a half years, and made over 500 million revolutions, delivering nearly 60,000 cu ft of air per minute.

The demand for continuous ventilation gave little chance for repairs or maintenance of the engines, and the practice developed of providing two engines, opposed to each other upon a single long bed, so that the connecting and eccentric rods could be changed over rapidly to allow either engine to drive to the crankshaft.

Since the engine was in a house, and except for the Waddell, the fan was enclosed, complete photographs of steam mine ventilation systems are rare. Electrically driven axial flow propellor type fans set in circular drifts are now widely used.

When mechanical ventilation was introduced in the 1860s, a variety of positive displacement machines were tried as well as fans. Among these was the Nixon type with two pistons each 30ft wide x 21ft high x 7ft stroke, worked by one engine, which was used at his Navigation Pit in Glamorgan. Another was the Struve, with circular air bells of 18ft diameter x 7ft stroke. These were the culmination of half a century of experiments. From a square-piston air pump by John Buddle in 1807 followed proposals for horizontal fans on vertical shafts in the 1820s, with Struve's type evolving from the 1840s. Brunton's horizontal fan also gave good results then. The small areas of colliery roads restricted the airflow but the safety of mechanical ventilation compared to the use of furnaces led to trials of almost every form of air displacer. So from vast eccentric drum displacers, Roots blowers, etc, the value of the fan was established, and soon fans with as great a variety of construction as the displacement exhausters were tried, ensuring safety, and even ventilating old gassy wastes. The spectacular open Waddell fan up to 50ft in diameter was very successful, as were the Guibal, Walker, and high-speed Schiele fans.

Waddell Fan Engines

53, 54   The Clay Cross Co, Morton Colliery, Derbyshire.   1965
J. R. Waddell, Llanelly, 1890. 30in x 4ft. Slide valve, non-condensing. 40rpm. 65lb/in².
Waddell was given the whole contract when Morton Colliery was converted to fan ventilation after 25 years of a furnace system. The engine was the plainest possible, with a Meyer cut-off valve, and ran continuously for 60 years, with only a couple of small repairs. The fan, designed to move 70,000 cubic feet per minute at 45rpm, was 40ft diameter and 1ft 6in wide at the rim. The photograph shows the curvature of the rotor blades by the line of rivets, as well as the simple construction of the Waddell fan from flat plates, rivets and angle iron.

55, 56   Chanters Colliery, Tyldesley, Lancs.   1965
Waddell fan 40ft diameter, Musgrave engine 28in x 4ft, 1896. 46rpm.
The fan rotor was fitted with an évasée rim to improve its performance to over 150,000 cubic feet per minute. It was also fitted with an extended shaft and crank for the high class Musgrave Corliss valve engine.

54

55

56

Walker Fan Engines

Walker Brothers of Pagefield Ironworks, Wigan, made a wide variety of colliery machinery, and a speciality was their 'Indestructible' fan. The largest ones, as at the Severn Tunnel and for Modderfontein Mines, South Africa, were direct driven, but most colliery fans were for higher speeds from rope drives.

57 Crumlin Navigation Colliery, South Wales.   1967
No 14099, 1908, fan 24ft diameter x 8ft wide, 300,000 cu ft per minute at 6in water gauge. Engine 15: 23½, 26 and 26in x 3ft 3in. Corliss valves.
Most Walker fan engines were horizontal cross-compound, the later ones with Corliss valves, and the two at Crumlin were the only triple-expansion slow-speed open type fan engines I knew. Very efficient, they were fitted with Whitehead's Corliss trip gear and Le Blanc condensers. One engine was a standby.

58 Dean and Chapter Colliery, Ferryhill, Co Durham.   1950
No 10572, 1902. Fan was standby, dimensions unknown. Engine about 18 and 32in x 3ft.
This ran at about 40rpm to develop about 250-300hp, probably non-condensing. It was the only Walker's open-type vertical fan engine I knew, again with Whitehead's trip gear.

Drift Mine and Haulage Engines

Drifts, ie following down the seam from the outcrop, are a very old form of mining. Raising the mineral was simpler, by a straight road and endless or single rope bringing the tubs to the bank. It needed a heavy pull at low speed provided by a geared drive, and a single or endless rope.

59 Smeaton Mine, near Dalkeith, East Lothian.   1948
Maker and date unknown, c1890. Called *Caledonia*. Tandem compound. About 18 and 28in x 3ft. Slide valves, condensing. 40rpm? 80lb/in^2.
Smeaton was a very old drift coal mine, worked out and closed by 1948. There were two other drift haulages by Inglis of Airdrie, but this tandem compound engine with endless rope, hauled up to 1,000 tons a day up over a mile of 1 in 3 slope. Fitted with a flywheel and condenser it ran continuously, the drams being clipped on to the rope. It was probably of local make, dating from the 1890s.

60 Morlais Colliery, near Llanelly. Drift, 600yd of 1 in 6.   1965
A. Barclay and Co, Kilmarnock, 1907. 15 and 24in x 2ft 6in. Slide valves, non-condensing. 80lb/in^2.
This probably began drift working to the Swansea Five Foot Seam in 1873, the date on the heading of the dip, and 1907 was probably the date of alterations under Thomas Williams of Llangennech. Most of the coal was later raised by the twin-cylinder Barclay engine of 1907 in the vertical shaft of the 5ft seam. This was very fast, working in 30-second cycles.

61 Warden Law, Bank Haulage for Hetton Colliery, Co Durham.   1951
Thomas Murray and Co, Chester-le-Street, 1836. 39in x 6ft 2in. Slide valve, non-condensing.
The Tyne was the main shipping point for the Durham collieries, and as the trade and collieries developed away from the river, it often meant crossing over the hills that lay between the pits and the Tyne. Haulages thus became a feature of the district, and Warden Law was typical. This beam engine raised five 12-ton trucks per journey over 760 yards of 1 in 19 incline at 8mph.

62 Bank Top Haulage, Burnhope Colliery, near Lanchester, Co Durham.   1950
Beam engine: Thomas Murray, c1845, 27in x 5ft.
Horizontal engine: maker unknown, c 1860s?, 24in x 5ft. 20rpm. 40lb/in^2.
Bank Top performed a similar service for Burnhope Colliery, but the pit developed, probably following the sinking of No 2 shaft about 1863-4. The tonnage was then too great for the beam engine, so a horizontal engine was coupled to the same crankpin. The beam engine had a slip eccentric, and the horizontal had link motion, movement being controlled from the horizontal engine. There were two rope drums, engaged by sliding the drum and shaft; it could thus work either incline.

63 A South Wales Colliery. Maker and date unknown.   1960
A typical late example of the haulages used in every colliery. Run on steam or compressed air, they performed any duty that a rope could do, and, made in small sections, could be placed in confined spaces. The variable gears gave high speed, or great power.

64 Lightmoor Colliery, Forest of Dean, Gloucestershire.   1932
Unknown make and date. Engine underground. 9in x 1ft 3in? Slide valve, exhaust steam to furnace ventilator.
Rapid movement of the coal to the shaft bottom was essential, and this engine, placed over the main haulage road, drove an endless rope by a 5ft diameter pulley. The drams were clipped to the rope which ran at about 3mph. The underground haulages were steamed by two egg-ended boilers in the pit, almost certainly sent down as plates and built up *in situ*.

59

60

Blast Furnace Blowing Engines

The high temperature required to smelt metals from ores is achieved by forcing air through the furnace. The steam engine was soon applied to furnace blowing, and Boulton and Watt supplied twenty-four for this purpose from 1785 to the expiry of the patent in 1800, and Newcomen engines were also used. At first they were simple direct-acting units, similar to Cornish engines. Rotative engines driving the air blowing cylinders ('tubs' as they were termed) in various ways followed, and the sketches in the technical notes illustrate a few of these. The 'horn' beam with the crank end upswept was a feature met in blowing engines but not elsewhere, and beam-engine blowers were built up to the largest sizes. The Darby engine at Ebbw Vale was one, with a steam cylinder of 72in and blowing tub of 144in bore x 12ft stroke. It pumped 44,000 cu ft of air per minute at 20rpm, and except for the cylinders was largely built by the ironworks' staff.

Pumping air at low pressure seems a light duty, but working for many months nonstop, and the heavy parts, tried everything severely, and only continuous strengthening of parts that failed overcame breakdowns. The many different arrangements were attempts to match the variations in the steam effort, and the blowing load, but however directly connected, the stresses found any weak spot. Since a blast failure could ruin a furnace, ample spare capacity was usually provided. Thus in 1890 the Edgar Thompson ironworks at Pittsburg, USA, with nine furnaces had a total of twenty-six blowing engines steamed by 152 boilers. Most of the air tubs were 84in in diameter, and the works made 10,850 tons of iron per week. English furnaces were driven less hard, but spare blowing capacity was general.

The air and steam cylinders were later coupled together closely, so that the rotative motion had only to absorb the difference between the applied effort of steam, and opposing load of the blower tub. For many years, the engines were simple expansion only, but often two engines were coupled together, using steam at 40 to 60lb/in^2 from boilers fired by blast furnace gas, condensing where possible, although water shortage sometimes prevented this. With blast furnaces only there was usually plenty of gas and heat available, so the need for a massive reliable engine led to the retention of the beam engines, often as standby units, as at the Lilleshall Company in Shropshire. Some of these were later altered to give better performance. There was a unique nineteenth-century example of this at the Dowlais Ironworks in South Wales, where two single-cylinder beam blowers had been altered so that steam was used in three cylinders in succession in the two engines. In the first engine cylinders of 48in and 60in bore drove a single 145in tub, and the exhaust steam then passed to the 60in condensing cylinder of the other engine, to drive a 108in tub.

The plain horizontal and vertical engines with the steam and air cylinders coupled directly, and with a crank and flywheel system for steady running, developed from the mid-nineteenth century. Again every disposition of the parts was tried out, and the examples show them to be neat and attractive. One was the long crosshead form, which was widely used in America, where the hard driving of the furnaces required large blowing capacity, and the engines noted at the Edgar Thompson steelworks were long crosshead units.

Economy in the use of once plentiful blast-furnace gas became necessary early in the twentieth century, when improved furnace techniques reduced the amount of gas available, and heavier demands upon the blowing engines required more power. The evolution of the combined iron and steelworks also required ever more rolling-mill power, needing fuel economy everywhere to reduce the coal used in the boiler plants. Compound condensing blowing engines of advanced design, with improved tubs and valves, played their part in this. Steam turbine blowers were introduced early in this century. They used less steam, were smaller, and needed little maintenance, all with a notable absence of the thuds and bumps that characterised the tub blower.

Continuous efforts to improve the efficiency of the air tubs led to the use of cuff-type poppet valves, piston valves in tubular piston rods, and other devices at the expense of driving mechanisms, even moving cylinders (Slick's patent), which gave very large port areas by sliding the heavy cylinders. Each was an improvement in some respect, and later, very light plate-valves met the need to work at higher speeds for gas engine drives, and finally the turbo blower which was economical of space, steam and maintenance costs, came to the fore and was occasionally electrically driven.

Beam Blowing Engines

The direct acting beam blower was simply an air pump in which the air cylinder (or 'tub' as it was called) replaced the pump. Some were Newcomen engines, and Boulton and Watt made a number by 1800. The cylinder and tub were at the opposite ends of the beam. Rotative engines soon followed and with these the cylinder, air tub and crankshaft were placed in every possible relationship, to reduce the heavy strains and gain steady working. The 'horn' type beam (Plate 67) was almost exclusively a beam blower feature, often giving a longer stroke to the cranks than to the cylinders and tub.

65   Westbury Iron Works, Westbury, Wiltshire (1857 on house).   1930
Maker uncertain, possibly Harris, Darlington (on roof tank), 1857. 40in x 7ft with Musgrave HP 40in x 3ft 6in. Air tub 84in x 7ft.
This ironworks had a chequered history and the plant indicated various injections of new capital. Thus higher pressure boilers were put in in 1873 when Musgraves compounded the engine. The vertical engines came later (Plate 69), but the works closed about 1900. *Engineering* (1903) notes that 'the Westbury Ironworks closed for 3 years is to be opened by a new company'. The installation of a (steam) turbine blower and horizontal four-cylinder gas engine, followed later, again possibly with new boilers, of which there were seven, gas fired, in the derelict works in 1930. The steam and air cylinders were at the opposite ends of the beam, with the connecting rod near to the steam cylinder (foreground). Beyond the cross entablature beam and supporting columns is the 1873 high-pressure cylinder, with the air tub farthest away.

66-67   G. and R. Thomas, Hatherton Furnaces, Bloxwich, Staffs.   1951
Maker and date unknown, possibly 1860s. Steam cylinder 48in x 9ft 6in. Air tub 104in x 9ft 6in. 60lb/in^2?
The works were not latterly making iron, and the engine was disused. It was a large and modern example of the beam blower, that had had a long and hard life over some 70 years, and the cracks in the entablature suggest subsidence. Care had been taken to make it neat and attractive. The blowing tub was 9ft in outside diameter by 11ft high. The wooden connecting rod was 19in x 10in full section, and 27ft long, had a half-lap joint near the middle. Plate 67 emphasises the general neatness of the raised, or horn, end beam, and light attractive railings. The flywheel was about 22ft in diameter, and the crankshaft was about 3ft beyond the end of the beam with the connecting rod sloping outwards.

68   Goldendale Ironworks, Tunstall, Stoke-on-Trent.   1950
Maker unknown, originally 1833? About 45 and 48in x 7ft 6in. Cornish valves. Four air tubs 60 to 90in.
The engines were as remarkable as the works, with one open-top furnace making cold-blast iron for special orders in 1950, and the others warm-blast furnaces. The first engine, on the left, was probably built in 1833 to work direct, and came to Goldendale in 1844. The other one was installed later, again as a direct non-rotative unit, and much of it may have been made in the works, as parts had 'Goldendale' moulded in. They worked independent direct, until in 1860-70 they were made rotative by the addition of front beams projecting through the wall, and coupled to a single 16in crankshaft and the 24ft flywheel. They ran like this until in the 1930s the crankshaft was split, and the small flywheel added for the left-hand engine. The last change in the 1940s was the installation of a steam-turbine blower, with the beam engines as standby. There were four air tubs eventually. In every way it was a delightful piece of engineering.

7

8

Other Blowing Engines

The beam blower was reliable but, being slow, was large and costly, needing an expensive house. As horizontal and vertical engines were developed they were applied to blowing, and almost every possible linkage of steam and blowing cylinder was tried to secure even blast and avoid breakdown. The long crosshead types and others with direct connnection of the steam end and tub, both horizontal and vertical, were tried out. The long crosshead engine (Plates 70 and 71) was widely used, especially for the very hard driven American furnaces.

69   Westbury Ironworks, Westbury, Wiltshire   1930
Galloways, Manchester, c1875. 38in x 5ft. Piston valves, condensing. 80in air tubs. Galloways built this design in the 1870s, and they were about 25ft high. They were fitted with separate expansion valves, and steam balance pistons to even the loading. The steam turbine blower just seen at the left was the main blower in later years and there was a horizontal four-cylinder oil or gas engined generating set.

70   Staveley Ironworks, Staveley, Derbyshire.   1962
Galloways, Manchester. Numbers 1-4 1907, No 5 later. 36in x 5ft. Piston valves. Air tubs 90in, altered to 110in later.
The first four engines were new from Galloways, the other second-hand from another works. They were built to run condensing, to blow 19,000 cu ft per minute each at 44rpm, but 110in tubs were later put in, to blow 21,500 cu ft per minute at 33rpm. These were pure long crosshead engines with crosshead guides outside the frame, and twin 18ft flywheels for each. They were over 30ft high. Escher Wyss steam turbo blowers, for steam at $350lb/in^2$, replaced them in the 1940s, but the Galloway engines were still standby in the 1960s. The plate shows the stiff cross braces in the frame, and the mechanically operated air valves in the later tubs.

71   Kettering Ironworks, Kettering, Northants.   1960
Galloways, Manchester, 1909. 36in x 5ft. Two piston valves. Air tubs 100in x 5ft.
This shows details of Galloways long crosshead design. The flywheels were 18ft diameter, and the engine was 34ft high to the top of the air tub. The side frames were single castings over 22ft high. There was a cam-driven variable cut-off piston steam inlet valves, with a separate piston-type exhaust valve. The mechanical air valves for the tub were driven by an eccentric on the crankshaft. These replaced the original leather flap valves in 1912.

72, 73 Kettering Ironworks, Nos 1 and 2 engines. 1960
Kitson and Co, Leeds, 1878 and 1881. 40in x 4ft 6in. Originally slide valves. 84in air tubs.
In the engine (Plate 72) the slide valve had been replaced by a piston valve, otherwise the engines were as built. The air tubs were attached directly to the frames, with the steam cylinders on top, totalling over 28ft high. A massive cross stay stiffened the frames, and carried the four-bar crosshead guides. There was also a Lilleshall engine of 1894 similar to the Kitson, with a 44in steam cylinder and 100in air tub, a double-throw crank and two flywheels. They were all non-condensing. From 1931 new boilers (150lb/in^2) drove a 250kW turbine exhausting to the old 70lb/in^2 range.

The Southwark or quarter crank blower had the steam cylinder and air tub separate, driven by cranks at 90°, and with a flywheel between. The air valves were of the grid type, with very short travel, driven by eccentrics on the crankshaft. The shield-shaped plaque on the front and the circular frames were other Southwark features.

74   Briton Ferry Ironworks, South Wales.   1958
Richardson Westgarth, Middlesbrough. No 343: 1909, No 301: 1906. Steam cylinder 42in x 5ft. Corliss valves. Air tubs 72in.
The works gate was dated 1843 and No 343 was bought new to replace a beam engine. No 301 was bought from the Blaina Ironworks when they closed about 1916. Typical Southwark engines, they were 28ft high with 21ft flywheels and ran non-condensing. No 301 was the standby engine; it was fitted with a condenser which was not used at Briton Ferry.

75   Appleby Frodingham Ironworks, Scunthorpe, Lincs.   1959
Richardson Westgarth, Middlesbrough, No 236, 1904. Steam cylinders 42in and 84in x 5ft. Corliss valves. Air tubs 84in. 35rpm. 160lb/in^2.
Although these engines were quite separate mechanically, they worked as a pair, compound, to develop about 1,800hp. The flywheels were about 18ft in diameter, and, standing some 30ft high, they were impressive engines.

76   Gjers' Mills and Co, Ayresome Ironworks, Middlesbrough.   1961
Cochrane, Grove & Co, 1870. 40in x 4ft. Piston valves. 90in tubs. 50lb/in^2.
These engines were installed when the works were built, and although quite separate they were tied together with turned circular stays. There were four bearings to each engine — one inside and one outside of each flywheel, and as the photograph shows they were very steady when working, the exposure being one minute. The steam cylinders were at the top. A similar engine made by Yates and Thom was installed in the 1890s. In 1872 one engine supplied 11,500 cu ft of air per minute to two furnaces which made 750 tons of iron per week.

77   Brymbo Iron and Steel Works, Brymbo, North Wales.   1963
Richardson Westgarth, 1942. Turbine blower. Capacity 19,050 cu ft per min at 4,660rpm.
This was bought second-hand about 1966 to replace a vertical twin-cylinder Davy Brothers blower, which had run for nearly 90 years. Without a standby, it ran up to three years non-stop, and was over 30ft high by 15ft wide. The turbine was very small in comparison, as the attendant behind shows, and only about 16ft long overall.

Rolling Mill Engines

Rolling was the final stage in making metal suitable for everyday use, by passing it through power-driven rolls of various contours. There were three main types: the continuous and the three-high mills which revolved in one direction, and the reversing type which reversed each time that the metal went through the rolls (each pass). Continuous mills had large flywheels which stored energy and would pull through loads far exceeding the power of the engines. The reversing mill, with no flywheel, required an engine to meet all loads, but occasionally a flywheel engine drove a reversing train through gearing and clutches. In the three-high mill, the metal moved in opposite directions between the lower two and the upper two rolls.

The power needed varied with the product. The wrought iron trade was well served by engines of 300 to 500hp, with flywheels on the fast intermediate shafts (Plate 88), since the small workpieces could be handled by the men. The steel trade working heavy ingots needed up to 25,000 peak hp with mechanical handling everywhere. To make three long rails from a 4 ton ingot in 3 minutes, required three engines similar to Plates 78-82, each developing up to 8,000hp for the cogging, roughing and finishing stages. In the wrought iron trade, steam of $30lb/in^2$ was made by boilers using waste heat from the furnaces, and the simple reliable beam engines remained unaltered. When, however, a works changed from wrought iron to working steel billets bought in, there was little waste heat, and fuel had to be economised. The Bromford Ironworks at West Bromwich illustrates this. From 1854 to 1888, making wrought iron, the mills were driven by such an engine as Plate 88, with a 48in x 8ft cylinder using steam at $30lb/in^2$ from waste-heat boilers. New management then realised that steel ingots, bought in, must replace the wrought iron made there, and in 1898 new boilers for $120lb/in^2$ were installed and the old engine made compound by adding a cylinder on the front wall near the connecting rod. The engine was then able to work steel. Ten years later, another mill and triple-expansion engine with boilers for $200lb/in^2$ was installed; as it was still sound the old engine was made triple expansion by adding a further 24in cylinder near to the original 48in one. Steel trade engines therefore ranged from the re-rollers of 250hp to the large girder mill engines of 25,000hp.

There was little room for experiments, or costly delays, but new ideas were tried out. Thus, a very large Brotherhood three-cylinder radial engine was tried at the Cyclops Steelworks in Manchester in 1872. It did not do well, but a 750hp Belliss and Morcom enclosed engine and rope drive of 1913 did very well at a Tipton rolling mill. Another unusual layout was the mixed-pressure turbine and gear drive installed in 1905 at the Calderbank Steelworks. It developed 750hp but with the 90ton flywheel carried a 2,000hp peak loads on the three-high plate mill. At least two Manhattan-type engines served in metal rolling: one at the Rhymney Ironworks in South Wales was a vertical engine converted by fitting a complete horizontal engine driving on to the same crank pin. Both cylinders were 60in x 4ft with piston valves and Joy's valve gear; it was later converted into a compound engine. The other for the Wishaw works in Scotland was built as a Manhattan engine by Lamberton about 1907. The cylinders were 42 and 72in x 4ft with piston valves driven by Allan's valve gear.

The steam rolling mill in Britain reached its peak in the first quarter of this century, when several vertical three-crank engines of 12,000hp and one of 18,000hp were built by Davy Brothers and many large horizontal engines by Lambertons, Markhams and Galloways. As an example of the massive build of rolling mill engines a Hick Hargreaves 1,500hp six-mill engine, 32 and 65in x 6ft, $160lb/in^2$, 37rpm, weighed 212 tons without the flywheel, which was 30ft in diameter and 100 tons in weight. The last large rolling-mill engines were probably those built by Erhardt and Sehmer in 1947 and 1953-54, and the most powerful ever built was a 30,000hp five-crank, fully-enclosed high-speed set with hydraulic valve gear from GHH of Duisburg in the 1930s.

In 1919 the Mesta Company of Pittsburg built a 25,000hp twin-tandem engine for the Lukens Steel Company of Pennsylvania. The piston-valve cylinders were two each of 46 and 70in bore x 5ft stroke, using steam at $225lb/in^2$. It weighed 625 tons and drove the 17ft plate mill by 2 to 1 reduction gears 4ft 6in wide.

Reversing Rolling Mill Engines

The Park Gate Iron and Steel Company, Rotherham, was a very large iron making complex which, later in the nineteenth century, converted much of its output into wrought iron on the site. By 1880 when the change to steel making began, they had ninety puddling furnaces in use, which were soon abandoned for steel making, again from their own pig iron. The early rolling-mill engines were doubtless similar to Plate 78, and 80-82 indicate that the change from the 1cwt puddler's ball to cast ingot of several tons was reflected as much in the engines as it was in the works, where massive but quiet furnaces superseded the army of puddlers who toiled to make the tough wrought iron. In later years the engines exhausted to low-pressure steam turbines.

78 The Cogging Mill.   1955
Davy Brothers, Sheffield, 1880s. 50in x 5ft. Piston valves. Geared down 2½ to 1 to rolls. 80lb/in^2.
Cogging was the first stage, ie breaking down an ingot from about 16in x 16in x 5ft to 8in x 8in x 18ft long. This required a short but very powerful draught, readily controlled as the ingot was short, increasing as it was reduced in area. The geared drive gave this, and eighteen passes through the rolls, of two, then three, four and five revolutions of the engine per pass, reversing each time, were made in 1¾ minutes, using up to 8,000hp. These were the working revolutions, there were some dead ones as the live rolls returned the slab to the rolls. The reversing gear was the Marshall single-eccentric type, and response was very quick.

79 The Cogging Mill, later engine.   1955
Scott and Hodgson, 1917. Three cylinders 40in x 4ft 6in. Piston valves.
This engine replaced the Davy Brothers engine (Plate 78) when the Shotton Company had to increase the power of their billet plant (Plate 85) by installing a 42in electrically-driven slab mill. This three-cylinder engine was very handy, and was geared down to the mill. Installed at Park Gate about 1957, extensive casing was then added to keep out mill dust.

80 The Roughing Mill.   1955
C. Markham and Co, Chesterfield, 1921. Direct drive to 24in-mill. Three cylinders 36in x 4ft 6in. Piston valves, Joy's valve gear.
This was installed new from Markhams, and coupled directly to the mill, the counterbalanced cranks allowed it to run up to 140rpm, and develop 8000hp with steam at 160lb/in^2. Joy's radial valve gear, driven from the connecting rods, drove piston valves on the top of the cylinders. It looked very light and flimsy at speed, but as with all Joy's valve gear gave splendid service with little trouble. The frames were very stiff, and with the top valves, absence of eccentrics and couplings, the cylinders were close together and the whole engine was extremely strong and rigid.

81 Finishing Mill.   1955
Galloways, Manchester, 1912. Twin tandem 2 x 38 and 57in x 4ft 6in. Piston valves.
This was supplied new to Colvilles, Glengarnock Steel Works, together with two three-cylinder engines in 1912. Colvilles converted their works to electric drive in 1933, and the steam sets were taken out. This engine lay idle and was bought by Park Gate in 1940; it was at work in 1941 and ran until Park Gate closed in the 1970s. Galloways, like Markhams, made a very stiff rigid framing designed from sheer experience, and this engine was little wider than the cylinders. As with all finishing mills, the service was very hard, the speed and long passes being very demanding. Hick Hargreaves of Bolton supplied new cylinders, pistons and piston rods for this engine in 1950.

78

79

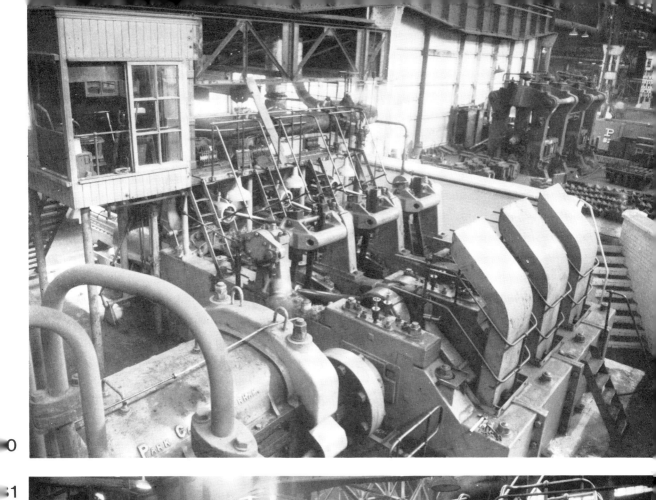

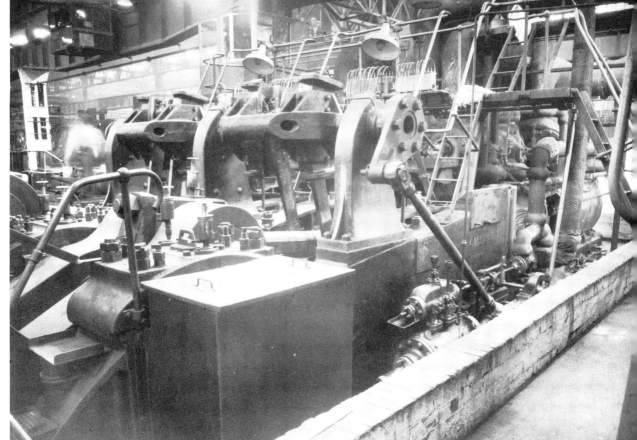

82   Park Gate Iron and Steel Works, The Billet Mill.   1955
Lamberton, Coatbridge, 1911. Three cylinders, 40in x 4ft 0in. Piston valves. 160lb/in^2. 140rpm.
This was supplied new to the works, to run on steam from the turbine power-house boilers which powered the generators and blowers. The turbines also used the engines' exhaust steam. It comprised three virtually independent engines with flange couplings between each crank. Piston valves at the side were driven by Allan's link motion. Well built and maintained, it gave the high speeds and powers required; new main bearings and piston rings were fitted to it in 1958.

83   Bynea Steel Works, Loughor, near Swansea. Tinplate Bar Mill.   1957
Lamberton, Coatbridge, 1912. 6000hp. Three cylinders 36in x 4ft. Piston valves. 160lb/in^2. 140rpm.
Bynea works was especially built to provide soft steel bar for tinplate making, the bar for which had steadily increased in size in later years. Generally similar to Plate 82, differences were that there were no couplings between the cranks, the piston valves were on the right, and the cast-steel cranks were counterbalanced. This was the original crankshaft, weighing 43 tons, which, when a loose crank pin was repaired by Markhams in 1955-7 was put back in the engine. A 19-ton shaft, fitted in 1955, was far too light and was removed in 1957. There was no room for couplings between the bearings.

84   The English Steel Corporation, River Don Works, 48in Plate Mill.   1961
Davy Brothers, Sheffield, 1905. Three cylinders 40in x 4ft 6in. Piston valves, non-condensing. 160lb/in^2. Up to 120rpm.
This was Davy's late design with each cylinder and motion line separate, couplings between the cranks, and with stays from the centre to the two outer standards passing through holes cast in the inner frames. Joy's valve gear, although light in structure, was adequate for the valve drives. It was geared down 4 to 1, and could roll an ingot down to a plate 40ft long 13ft wide and 3in thick. Rolling a slab 24in thick to 3¾in thick took fifty passes through the rolls in 16 minutes, each pass needing up to ten revolutions of the engine.

85   John Summers and Co, Shotton Steel Works, North Wales.   1970
Scott and Hodgson, 1917.
This is the engine shown in Plate 79 in its original position. It was almost soundless when at work, having a skilled fitter who only worked in this engine room, correcting the slightest fault. The gearing and mill were replaced twice to take larger billets. Latterly it used steam at 160lb/in^2 from the works ring main, which was passed out from turbines taking steam at 480lb/in^2. The engine exhaust passed into the works low-pressure mains.

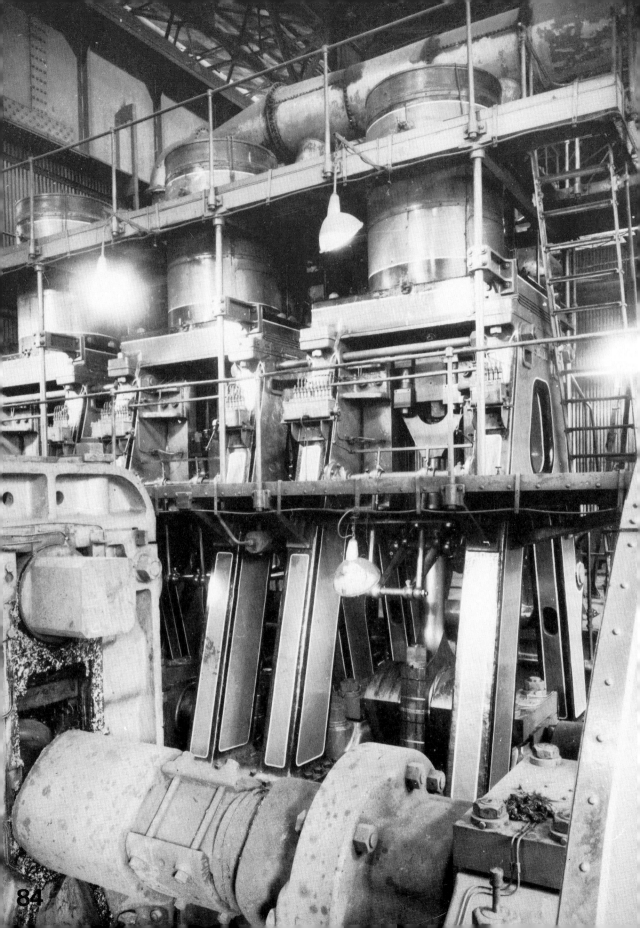

86, 87 The Sheepbridge Coal and Iron Co, The 20in Mill. 1970
Sheepbridge Workshops, 1910. 32in x 3ft 6in. Piston valves, non-condensing. 80lb/in². 600hp. 70rpm.
This engine was completed in the company workshops, although the main castings and crankshaft were probably prepared outside. The bed plates were single castings 24ft long and 4ft 6in wide. This was the last engine on the site, due to be replaced by an electrically driven mill. The rolling mill (Plate 87) was a two-stand general purpose set, sold to another concern who, however, could not use the engine, needing more power.

Midlands Ironworks Engines

The making of wrought iron by the puddling process developed early in the nineteenth century to become a great industry, especially in the Midlands and Yorkshire. Thus there were over 2,600 furnaces in the Midlands by 1870, and the Park Gate works in 1880 just prior to the change to steel production had ninety furnaces in one works. The output per furnace was small and they were often grouped in fours, usually with its own chimney and Rastrick flue type boiler. Utilising waste heat, these supplied steam to the rolling mill and other engines. The many chimneys and the tall casing around the boilers were long a part of the Midlands scene. It was a heavy trade, hard alike on men and machines, each growing in ruggedness to meet it, and anything that failed was replaced with a stronger part. Wheel arms were fitted with loose dovetails, filled in with timber and steel wedges. The engines were usually the simplest condensing single-beam type with Cornish valves and late cut-off, developing up to 500hp at 18-20rpm. The connecting rods were of timber, 20 or more feet long and strapped with iron plates each side. The drive was by gears to the second-motion shaft, on which the flywheel was fitted and ran at about 80rpm; the mill or finishing rolls were often driven from this shaft. The heavy forge or first rolls were also geared down from this. The cylinder was in a square brick house with the beam working on the wall. The steam pressures were low—30 to 50lb/in^2, well suited to the simple engines and long steam pipes from the scattered furnaces and waste heat boilers.

88 The Harts Hill Ironworks, Dudley, Staffs.   1938
Possibly Holcroft or Perry of Bilston, c1850s. 48in x 8ft. Cornish valves, condensing. About 400hp. 18rpm.
A typical Black Country mill showing the mill roll train to the right and the heavy forge train to the left. The valves were driven by a rocking shaft from a single eccentric on the crankshaft, with the steam valve of one end and the exhaust valve of the other, coupled to a single rod driven by a wiper from the rocking shaft below. The main driving gear was 16ft to a 4ft pinion on the shaft carrying the 20ft diameter flywheel. This drove the distant forge train by an 8ft pinion and 16ft gearwheel. The wooden connecting rod from the engine beam was about 20ft long with iron strapping on front and back, and the engine room wall was of brick 4ft thick. This engine ran for nearly a century.

89 Walkers' Rolling Mills, Walsall, Staffs.   1952
Maker unknown, c1880s. 42in x 5ft. Slide valve, condensing. About 450hp. 40rpm. 100lb/in^2.
The original Meyer cut-off valve had been replaced with a single valve, and the condenser was behind the cylinder with the air pump driven off the piston tail rod. The connecting rod drove on to a crank pin in the 8ft 6in pinion on the shaft carrying the 24ft flywheel, the pinion driving to the 10ft gearwheel and by the 18in square shaft to the two stands of rolls. On the other side of the flywheel, a similar pair of gears drove another two-stand mill with housing dated 1941. It was 44ft long from crank shaft to the rear of the air pump. Walkers were re-rollers only, so with no waste-heat boilers steam was supplied from five hand-fired Lancashire boilers, and this was probably the reason for adding the condenser, ie for economy of fuel.

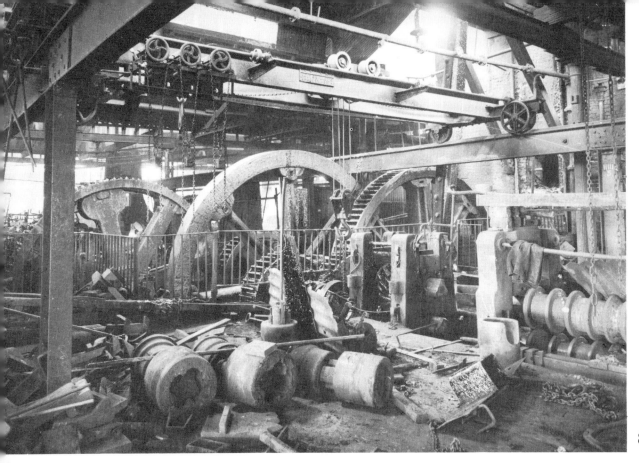

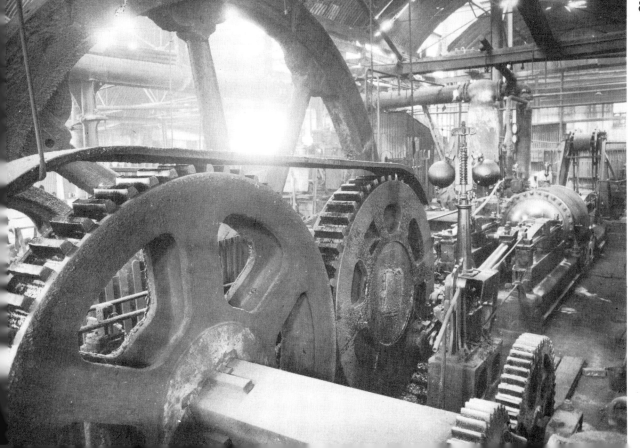

Continuous and Sheet Rolling Mill Engines

Tinplates and sheets were made from flat bar, pieces about 10 x 5 x ⅝in which could be returned over the top roll to the rollerman. The engine could thus run continuously, and so large flywheels could be fitted. Each 'mill' comprised two stands of rolls, one each for roughing and for finishing. The size of the bars, sheets and rolls grew and finally rolls of 26in diameter were used. The beam engines could only work one or two mills on small rolls and bars, but by 1880 larger engines drove up to four mills. The final stage with six mills, large sheets and rolls, first used at St David's Tinplate Works about 1909-10, required the use of flywheels up to 130 tons and 36ft diameter, and 1500hp.

90   The Glynhir Tinplate Co, Pontardulais, Glamorgan. 1958
J. Musgrave and Sons, 1911. Six mills. Cross compound. 28 and 50½in x 5ft. Corliss valves, condensing, 26rpm. 160lb/in^2.
This was identical to the first six-mill engine (at St David's) and was ordered in December 1911. Three mills each side were driven off a double-throw crankshaft with a 32ft diameter 120-ton flywheel. Hick Hargreaves bored out the exhaust ports and fitted new valves in 1956, and it was working superbly at the closure and scrapping in 1958.

91   The Gorse Galvanising Co, Dafen, near Llanelly. 1958
Hick Hargreaves and Co, 1912. Six mills. 33 and 64in x 6ft. Corliss valves. 1,200hp. 30rpm.
Very similar to Plate 90 in layout and drive, this was a sheet mill needing more power. The 33ft flywheel had ten arms and sectors with a rim 30in deep and 18in wide, and weighed over 130 tons. It ran until 1959.

92   John Lysaght, Orb Works, Newport. 1946
Galloways, Manchester, about 1912. Two tandems. 34 and 64in x 6ft. Corliss valves. 150lb/in^2. Up to 1,500hp. 26rpm.
Lysaghts completely modernised their sheet works from 1912, with six mill sets. These two, in the middle, were the first of the new engines, with Corliss valves and condensing tandem cylinders driving mills to one side only. A battery of six Galloways boilers steamed these two engines.

93   John Lysaght, Orb Works, Newport. 1946
Galloways, Manchester, about 1922.   60in x 6ft. Uniflow condensing. 1,500hp. 26rpm. 160lb/in^2.
When the next stage of the reorganisation was reached, the uniflow engine had proved its value, so it was adopted for the last two six-mill sets. They were two of the largest uniflows built, the first with radial and the last with Pillings oil valve gear. The flywheels, cast in South Wales, weighed over 150 tons each, were 32ft in diameter and fitted on a 31in square section of the 26in crankshaft. The ten arms were fitted to the rim by wood wedged dovetails. Each of the uniflow engines (one each end) had its own battery of five Galloway boilers. The steam plant was scrapped about 1950.

94   Partridge Jones and John Paton, Pontnewynydd. 1958
Hick Hargreaves and Co, Bolton, 1910. Two tandems. 35 and 65in x 6ft. Corliss valves. 1,500hp. 32rpm. 150lb/in^2.
These were said to have been arranged to each drive eight mills off one side. They were certainly very powerful tandem compounds, with 32ft-flywheels of about 120 tons each.

95   The Pontardawe Alloy Co, Pontardawe, South Wales. 1958
Cole, Marchent and Morley, Bradford, 1919. Six mills? 24 and 44in x 4ft. Drop piston valves, condensing. 1,200hp. 160lb/in^2. 90rpm.
This drove the mills by thirty ropes to a 30ft Musgrave flywheel on the mill shaft, and was fitted with a Musgraves Radojet condenser. It was well kept and gave good service, but it was idle by 1958 and went in the general scrapping of the hand mills when strip rolling began in the 1960s. Again, it was a standard engine adopted for slow rolling.

90

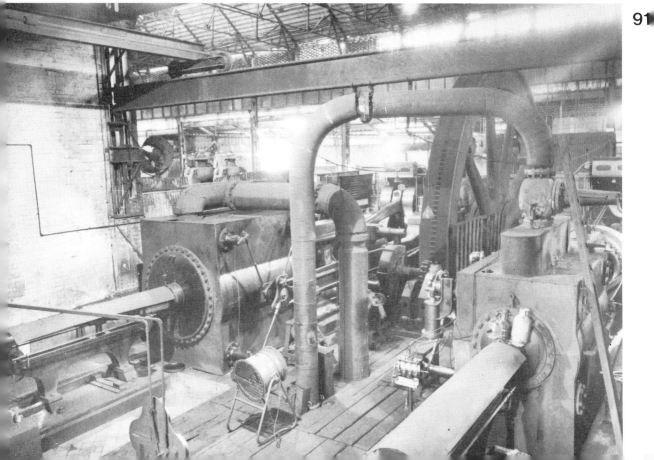

91

94

95

Tinplate Hot-Rolling Mill Engines

Tinplate required two rolling stages: the hot mill reduced it from the short thick tinplate 'bar' to the required thinness and width; and cold rolling planished the surfaces. The slightness of the thin plates was deceptive, as they required great power in the mills to reduce the thickness with the least number of passes and heatings. In the later stages, the plates were 'doubled', ie several going through at each pass. For many years, a beam engine or waterwheel drove one or two mills of two stands of rolls each. To drive four mills needed such engines as Plate 97 with a 40-ton flywheel, but when the six-mill sets were introduced about 1910-14, flywheels of 130 tons and 36ft diameter were required, and their transport to the site was a great problem. For many years, the mills were driven directly by the engines at 26rpm, with a 26in and 50in x 4ft engine driving four mills. The six-mill engines of the 1910-14 period were a little larger but, using steam at 160lb/in^2 or more and 130-ton flywheels, were much more powerful. These plates illustrate the various phases, as well as rope and geared drives. Another factor in load growth was the continuous increase in tinplate bar and sheet sizes.

96   The Kidwelly Tinplate Co, Kidwelly, Carmarthen.   1958
Edwin Foden, Sandbach. About 24 and 43in x 3ft 6in. Piston valves, condensing. 500hp, 26rpm.
Tinplate manufacture had long ceased here and little was known of the plant at the site. There were two engines, one of them an Edwin Foden, one of four 43in single-cylinder verticals he supplied to the trade probably in the 1880s. The other was similar but without a name. The plant was re-organised, and new capital provided boilers in 1892, when probably the engines were made tandem compound (the high pressure cylinder is on top, in the roof). The engine shown drove mills on either side, three or four each. The bar shears, and gearing for slow roll turning were a part of most tinplate hot mills.

97   The Baglan Bay Tinplate Co, Briton Ferry.
E. Foden, Son and Co, Sandbach, four mills, 1880s? 26 and 52in x 4ft. Piston and drop valves, condensing. 75lb/in^2.
This shows the standard Foden four-mill engine as supplied with piston valves, jet condenser, and cranks at 180°. Foden's engines were sound and well made, and with good maintenance there were few signs of failure, despite the heavy loadings. This engine had been fitted with a new HP cylinder with drop valves (Price's patent of 1895). The slow-turning gear and dog clutch used for roll turning at about 1rpm can be seen between the flywheel arms. The drop valve gear was very effective, often not admitting any steam on light load. The 25ft flywheel weighed about 50 tons.

99   Bryngwyn Steel Sheet Works, Gorseinon, Glamorgan.   1956
Galloways, 1909. 27 and 50in x 4ft. Corliss and piston valves. 1,500hp. 60rpm. 160lb/in^2.
This was a very powerful plant, geared down about 2½ to 1 to drive four sheet mills off each side. The works was started in 1899 and this plant was put in for six mills, and others added to it when in 1919 it came under the Grovesend Company. Very heavily worked, it had a new HP cylinder, crankshaft and flywheel after 1948, but closed in 1958 when strip production replaced hand mills in South Wales.

100   Baglan Bay Tinplate Co, Briton Ferry.   1946
Galloways, Manchester, 1930. 35in x 2ft 10in. Uniflow. About 1,200hp. 139rpm. 160lb/in^2.
This was probably the last engine that Galloway made, as well as the last large steam rolling-mill drive for South Wales. Fitted with Pilling's oil valve gear, it was very quiet and efficient, and geared down about 5 to 1 to the rolls. Three hand-fired Lancashire boilers steamed the Foden (Plate 97), uniflow and cold-roll engines.

98 St David's Tinplate Works, Loughor, near Swansea.   1956
Daniel Adamson and Co, 1903. Twenty-five-rope drive, five mills. About 21 and 36in x 3ft 6in. Piston and slide valves, condensing. 80rpm.
This was the first rope-driven tinplate hot mill, replacing the beam engine which could only drive three mills. The HP cylinder had a variable-travel governor-controlled flat cut-off outside of the main piston valve. This plant was replaced in 1909 by the first six-mill tinplate unit in South Wales.   The small engine at the right was possibly for slow turning, but in 1956 only the engines survived of the 1903 plant.

Tinplate Cold-Rolling Mill Engines

This was the finishing stage in making the basic plate for tinning. It consisted of passing each plate through two or three stands of rolls in succession to secure a high surface finish. It required much less power than hot rolling since it did not reduce the thickness, and the whole cold roll unit was driven by one engine of 200 to 400hp. The drive was almost always by ropes or gearing.

101   Baglan Bay Tinplate Co, Briton Ferry.   1946
Galloways, 1902. About 13 and 22in x 2ft 9in. Piston valves. 215hp. 95rpm. 120lb/in².
Galloway's superposed engines were extensively used for cold roll drives, and at Baglan Bay drove two sets of four stands of rolls each by gear-wheels of 5ft and 9ft 6in diameter, all off one side of the engine. It was Galloway's standard engine with the jet condenser at the rear of the LP cylinder, slipper and single bar guides, and 12ft flywheel. An interesting feature was the crankshaft governor which, housed in the circular case beside the crankshaft bearing, varied the power by altering the travel of the HP valve.

102   St David's Tinplate Works, Loughor, near Llanelly.   1956
Cole, Marchent and Morley, 1914? About 17 and 35in x 3ft. Drop piston valves. 350hp? 84rpm. 160lb/in².
A good example of the high-class engines installed in the last great period of hand tinplate mill reorganisation, and another standard engine adopted in the tinplate trade. The flywheel was 17ft 6in diameter, and three lines of rolls were driven by twenty-four ropes from the 9ft 6in pulley to 14ft 6in mill drive wheels. The jet condenser is seen behind the LP cylinder. It had worked hard with great economy.

103   Abercarn Works, Abercarn   1948
Hick Hargreaves and Co, c1925. 26in x 3ft. Uniflow. 500hp. 130rpm 160lb/in².
Another example of tinplate engine modernisation, this was a high power drive to three sets in tandem of three mills each. The massive flywheels seen in Plates 102 and 103 show how great the load could be, even in the lighter side of tinplate making. An interesting feature of this engine was that the lay shaft was on the outside of the engine probably due to the rope drive to a Hick Breguet condenser below. The framing of the engine was especially neat and strong.

104   John Player and Co, Clydach, Glamorgan.   1949
Hick Hargreaves and Co, Works ref number 1414, c1914? 24in x 3ft. Tabourin's Uniflow. 250hp. About 50rpm. 160lb/in².
This was installed second-hand probably in the 1920s. It was very unusual in that there were no tandem roll stands — simply four sets on each side, driven directly from the crankshaft. It was also a Tabourin's patent uniflow, with the exhaust controlled by a vertical piston valve at the side, operated by an eccentric on the side layshaft. There was a tail-rod operated condenser and a single flywheel. It was the only Tabourin type engine I met.

01

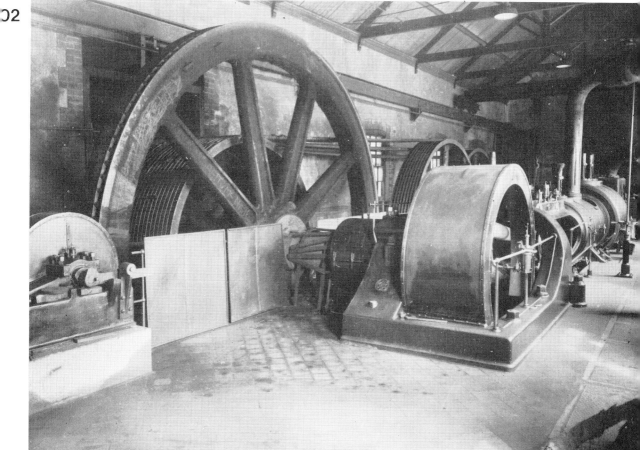

02

103

104

105 Melingriffith Tinplate Co, Whitchurch, Cardiff.
Summers and Scott, Gloucester, 1906? 13 and 24in x 3ft. Piston valves. About 150hp each. 80rpm. 140lb/in².
Melingriffiths cold-rolling mills were unusual in that the engines were vertical, and there were two completely independent sets. The engines were single crank tandem-compound condensing, with Hackworth valve gear, and condenser air pumps driven off the mill shafts. The mills were four tandem sets of two stands to each engine, driven off one side by 2 to 1 gearing. The engines were 23ft high, and the flywheels were 11ft in diameter. One engine was badly damaged when it ran away in 1926, and was repaired by Scott and Hodgson of Guide Bridge, Cheshire. The previous single-cylinder engine drove twelve cold rolls.

Engines For Metal Working— General Processes

A brief note on engines for working upon, rather than producing steel. They indicate design variations to meet the demands of the processes, as well as the drives to machines producing tubes, tyres, rods and wire.

106   Stewart and Lloyds Tube Works, Newport, Monmouthshire   1956
Cole, Marchent and Morley, 1917. Two twin-tandem engines. 23 and 44in x 4ft each side. Drop piston valves. 2,500hp. 85rpm.
This was a new works started in 1917, with an identical engine to each of the two main processes of piercing and rolling. Supplied with steam at 180lb/in^2 and 600^{0}F, they used 10½lb of steam per HP hour. The flywheels each 18ft 6in diameter x 8ft 9in wide were grooved for forty-two 2in driving ropes and weighed 120 tons for one and 80 tons for the other, driving to pulleys of similar weights. They were standard Cole, Marchent and Morley engines.

107   Taylors Forge, The Tyre Mill, Trafford Park, Manchester.   1968
Lamberton, Coatbridge, 1923. Three compound engines. 28, 34 and 34in x 3ft. Piston valves. 2,000hp max. 120rpm. 140lb/in^2.
The three mills made railway tyres, each engine driving one of the three stages between the blank pierced in the steam hammer and the finished tyre. They ran non-condensing, exhausting to feed-water heaters, and much of the steam was made by waste-heat boilers over the furnaces. The steam piercing-hammer was to the right (Plate 117) with the engines, each about 40ft apart driving the roughing, forming and finishing mills in that order to produce a railway wagon tyre in 4 minutes from leaving the furnace, with forty-eight, thirty-five and thirty revolutions of the engines.

108, 109   T. Whitehead and Co, Rod Mill, Scunthorpe, Lincs.   1959
Fullerton, Hodgart and Barclay, No 571, 1907. 32 and 50in x 4ft. 1,500 peak hp. 160lb/in^2.
This was a separate unit in the RTB complex, producing ¼in or ⅜in rod from 2in x 2in billets. These were reduced to 1in rounds in a six-stage reduction roll unit driven off the crankshaft, and finished in looping mills. The engine was the makers' standard condensing design. Plate 109 shows the drive from the 20ft flywheel 5ft wide, by the 56in belt, over two idlers, to pulleys on the three mill shafts, the centre lower one only being used for ¼in rod. The whole plant was moved from Tredegar in 1934, and it was intended to fit an electric drive in 1959, but retain the belt drive.

110, 111   Winterbottoms, Oxspring Wire Mill, Thurgoland, South Yorkshire   1948
Butterworth, Albert Works, Manchester, 1870s. About 18in x 2ft 6in. Slide valve, non-condensing. 30hp. 80lb/in^2. 90rpm.
This little plant, set in a country hillside, continued the reduction begun by Plates 108 and 109. The engine was the simplest possible, but the cylinder was a superb piece of foundry art, with steam inlet, cylinder, valve chest, and exhaust belt all in a single casting. The mill was on two floors with a waterwheel 14ft diameter x 5ft wide at the other end to the engine. This drove to the upper floors by gears and a belt with nine drawing blocks driven from a single shaft and bevel gears. The engine drove a similar series of blocks on the ground floor by gearing from the crankshaft. A belt, near to the engine, connected the upper and lower shafts, and there were four other blocks, at right angles, on the lower floor. There were no rumple or heavy blocks, but it would draw wire from about 8BWG to the finest sizes. The engine was overhauled in 1943 and all including the gearing ran very quietly.

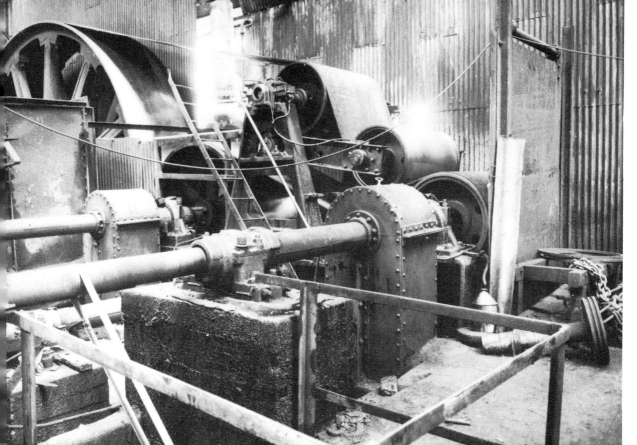

10

11

Works Services

The works services were most important, providing the mechanical aids that alone made the modern developments in speed and the weights handled possible. The power and rapidity of hydraulic systems varied from a 2in jack to a 6,000-ton press, and worked with certainty of action and response. Electricity was equally valuable and more versatile, and now compressed air is used in all departments.

112   Park Gate Iron and Steel Works, The Old Power Station, Rotherham   1965
C. Markham and Co, 1890s. Opposed ram pumps off tail rods. 20in x 3ft 6in. Slide valves, non-condensing. 40rpm. 80lb/in^2.
A Berry and a Markham pump, both similar in design, supplied water at 900 and 1200lb/in^2 for the hydraulic manipulators, turners, etc. These were essential when steel ingots up to 5 tons in weight replaced the 1cwt ball from the puddling furnaces (Park Gate had ninety of these in the 1870s). The hydraulically operated fingers in the live roll trains placed the workpiece into the required roll groove with speed and precision that was as amazing as the skills of the rollerman in his pulpit. The pump unit, placed upon a two-piece bed 30ft long, kept the stresses self-contained, and the photograph, taken over thirty or more revolutions of the engine, emphasises the solidity of the whole. The flywheel was 11ft in diameter, and the photograph shows the every day condition of a fine works.

113   Park Gate Works, The Old Power Station, Rotherham.   1965
Belliss and Morcom, No 4278, 1909. Three-cylinder compound. 28, 29 and 29in x 1ft 5in. Piston valves. 1080hp. 300rpm. 120lb/in^2.
The convenience of electricity for lighting and power led to its early use in the works, and by 1909 required this set for supply, with a 750kW DC Westinghouse generator. Designed to run non-condensing from the works steam system, the three-crank compound engine gave even turning and good economy at 120lb/in^2. The use of piston tail rods was interesting.

114   Park Gate, The Old Power Station, Rotherham.
Taken from the upper platform of the Belliss and Morcom engine, the switchboard on the left indicates how the DC network had grown. The two rotary converter sets in the foreground had been added by 1915. In the distance, are the two hydraulic pumping sets (see Plate 112) the Markham engine in full use with the standby Berry set to the right, behind the small AC switchboard.

115   Abercarn Tinplate Works, Abercarn.   1950
J. Musgrave and Son, 1911. 23½in x 2ft 9in. Uniflow, condensing. 350hp. 250kW.
An English uniflow engine built under Stumpf's licence, this supplied current to the works for half a century. It had a reciprocating camshaft on top of the cylinder, operated by the crankshaft governor seen behind the railing. Still used regularly, it was relieved of weekend duties by the motor generator set in the distance.

116   John Bagnall and Co, Lea Brook Ironworks, Wednesbury, Staffs.   1947
James Davies, Wednesbury, c1850s. 18in x 3ft 6in. Slide valve, non-condensing. 25rpm. 30lb/in^2.
This, the last surviving beam-engine-driven roll-turning unit, was, despite its apparent antiquity, perfectly serviceable in the hands of the men used to it. The beam was 11ft long, and the flywheel 9ft and the large gears 8ft in diameter. Roll turning, to restore surfaces damaged or worn hollow in working, was performed very slowly, at 1 to 2rpm, to preserve the hard surface. The roll to the right was driven through a train of seven gearwheels, and that to the left through six gearwheels, so that each roll turned towards the operator in finishing. The belts drove the indispensable grindstone that dressed the broad tools used to true the hard chilled roll surfaces.

112

113

114

115

16

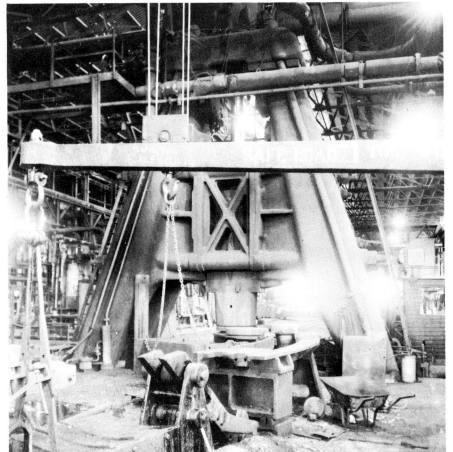

17

117   Taylor's Forge, Trafford Park, Manchester.   1968
The Morgan Engineering Co, No 1176, c1923. About 42in x 3ft 6in. Piston valve, about 40ft high. This steam hammer performed the initial forging process for making railway tyres, taking a thick blank, giving three blows for the first and the second stages to spread the blank to the full diameter. The piercing die (seen in the curved race near the anvil) was then lifted on top of the blank, and two or three hammer blows half punched the centre; the blank was turned over and pierced right through. The red hot blank was manipulated by hydraulic fingers at every stage, and the hammer produced a thin pierced ring from a thick blank in one minute, to be finished into a rail tyre in the three mills of Plate 107.

Boilers

Without the boilers there would have been no steam power, the history of which was one of generating and utilizing ever increasing pressures and temperatures. Many of my subjects were steamed by the simplest types, able to work with poor fuel and water and little attention. The plain egg-ended form was widely used in collieries and ironworks, and the Cornish and Lancashire internally fired boilers were used from the mid-nineteenth century. The works making wrought iron also used many Rastrick boilers. At first they were roughly made with hand-shaped plates often pulled into place when the punched holes did not meet correctly for rivetting up. The egg-end boiler contained a large volume of water relative to its evaporation and steamed steadily on the poorest fuels. They were also easy to clean — as indeed they had to be with the bad water of many collieries and ironworks. They were supported on stone or brick settings, usually with the gases passing along the lower half of the shell to the chimney (the flash flue as it was called), but occasionally the gases went around the sides (wheel flues), but flash settings with a simple wall between each shell gave closer placing with little heat loss. The ends were attractive double-curved plates usually attached to a circular end-plate, but in the North East were often finished with a single plate across the centre. They were 4½ to 6ft in diameter, up to 45ft long for collieries, but for ironworks with blast-furnace gas firing they were longer (60-70ft) and occasionally 80 to 90ft long, but the great lengths were of little value, and probably stressed rather than aided the boiler as a whole. They were suspended in various ways: by cast-iron lugs resting on the side walls, from cast-iron arches with vertical bolts to lugs on the shell, and occasionally by girders resting on vertical ones in the setting walls. The evaporation was fair, but few reached the 2,400lb per hour evaporated by the 37ft by 6ft boilers at Rainton colliery in the North East, but these were fired by Jucke's chain-grate stokers, giving a constant flame with the fire doors rarely opened. The simple externally-fired boilers had a poor safety record, and many egg-end boilers failed from seam rips over the fire bridge usually due to progressive cracking between the rivet holes. This led to shells blowing apart to become projectiles of incredible destructiveness from the sudden release of the energy stored in the hot water. This occurred even at well kept plants as at the Warrenby Ironworks near Redcar in 1895, when eleven of their range of fifteen boilers exploded. Ten of them failed from seam rips, and parts of one went 100 yards in one direction and 250 yards in the other. The chimney was demolished in the shambles, but not the blast furnaces; sadly eleven lives were lost and many injuries resulted. Mr E.B. Marten, reporting upon boiler explosions in 1870 noted that in the previous four years, ten explosions had occurred in egg-ended boilers, as many as in five times that number of Cornish ones insured. However, one expects the USA to go one better as it did at the Henry Clay mine in Pennsylvania in October 1894. Fired by unsellable small anthracite coal, they had thirty-six boilers, 43ft long by 3ft diameter set in threes, attended by three men per shift. As often, the occurrence lacked logic as the first five sets blew up, the middle nine boilers were only displaced, while the outer four sets exploded. One 40ft length of shell buried itself completely in the culm heap, and only the vast heap of culm saved the hundred men and boys working at the breaker on the pit top, and the nearby town of Shamokin, from disaster. The shift boilermen were lost but no one else. Even so, things were not always what they seem, as in the explosion of the set of twenty-two Elephant externally-fired boilers at the Freidrichshutte ironworks in 1888. There were eighteen in steam and three men in charge, and not only the boilers but also the works was destroyed. They were fired by blast-furnace gas, and in the intensive investigation that followed it was found that unburned gas had accumulated in the flues of two boilers that were being changed over; these exploded and set up a shock that destroyed the entire works. It was noted during the enquiry, that of 155 explosions that occurred in twelve years, fifty-seven were of boilers of this three-drum type, but there were large numbers used in factories, whereas there were few egg-end boilers used in British mills. The explosions emphasise the energy which is stored in heated water, but even here there was a lighter side. Thus one canny lad noting how the ends were scattered over the landscape in explosions thought 'it's the ends that are the trouble — I'll make a boiler without them' and he did so by making a 63ft long boiler in a ring or annulus of 25ft outside diameter which, using the waste heat from six furnaces gave good service. At Linthorpe ironworks in 1873, No 11 boiler in a range of fifteen exploded from a seam rip, and the ends flew apart injuring no one, with little disturbance to the rest. I must conclude this aspect with an almost laughable example of a Rastrick boiler in an ironworks at Stoke in 1873 where, one of eight, its water supply failed and it became red hot and a plate split. The occurrence only came to light when the water supply restored itself, and filling the red-hot boiler, it ran from the split plate back into the furnace. Despite these tribulations, such boilers continued in use, and later ones did well, and Messrs M. and W. Grazebrook used egg-ends at Netherton up to the closure in the 1950s. Messrs Walmsley of Bolton also used Rastrick boilers as long as they made wrought iron, and they were the last in the world to do so, but it must be noted that the later boilers were well made in the best style. Some of Grazebrook's were thirty years old, and the latest, of 1926, were steel. From the persistent usage of two types that were notoriously unstable it might be assumed that ironworks were old-fashioned and stick-in-the-mud traditionalists, but this was by no means so. When the Howard water-tube boiler was developed in the 1860s the Lackenby ironworks adopted them to provide increased steam pressure to allow the blowing engines to be compounded, which was rarely done until the turn of the century. Five were installed to blow the two furnaces making 800 tons of iron weekly, and three more were installed a year later to supply the extra furnaces projected. However the failure of welds in the tubes of Howard boilers caused many to be replaced later, but the Barrow Company were still making and selling them in the 1880s.

The Rastrick boiler, designed to use the great amount of heat rejected in the puddling cycle had many good points, but made of thin plates and of large diameter and racked by irregular heating, the mystery is, not that some failures occurred, but that they were so few. Today, the idea of such vast pots of boiling water merrily sizzling away amid the army of puddlers and helpers would be viewed with alarm, yet there were several works like Shelton Ironworks which, in the 1870s with ninety-four puddling furnaces, had a dozen Rastrick boilers in a row, each served by four furnaces with eight to ten men. With some 8,000 puddling furnaces in use in 1872, there were large numbers of Rastrick boilers in use and most works never had explosions, but when they did occur, the damage, and with so many men working near, the loss of lives was appalling. At Wells works at Moxley, in 1868, the explosion of one of their four Rastrick boilers virtually demolished the works, with the loss of thirteen lives, and at the Birchills works in May 1880, the failure of No 4 of their four Rastrick boilers which worked with other types, threw the shell over 120 yards away. The resulting devastation was notable even for a Rastrick boiler, and the loss of twenty-five lives and many injured made it one of the worst disasters in the history of land boilers, but it was a case of working a doubtful boiler. I can find no full record of boilers and explosions, but it seems certain that there were 100,000 boilers working in Britain in the 1880s. Some were nursed, and others neglected, but they were the mainstay of industry and served well.

118   Frog Lane Colliery, Coalpit Heath, sawmill at Westerleigh, Gloucestershire   1920s
Egg-ended boiler, maker and date unknown, about 20lb/in². About 4ft 6in diameter x 25ft long, flash flue.

This site was almost certainly the old New Engine colliery sunk in the 1830s. The beam-engine winder was probably made by Acraman of Bristol, and was geared to the rope drum, and when the pit closed the engine was adopted to drive Coalpit Heath colliery sawmill for pit props and general timber. Placed beside the engine house, the boiler setting was largely of stone. The steam and water connections were at the back, followed by the safety valve, which had weights at both ends of the lever. The plates were about ⅜in thick with a strengthening ring around the manhole opening. Next was the water-level indicator which was unique in my experience for this type of boiler. It was of the usual form with a stone float inside the boiler, attached to a rod passing up through the shell to a chain over a pulley with a weight at the other end. The weight indicated the water level by its position, but the feature, unique for an egg-end boiler was that on the rod up to the chain and pulley was a stop which, if the water got either too high or too low gave warning by operating an alarm whistle via the curved levers.

119   Crawshay and Co, Lightmoor Colliery, Ruardean, Forest of Dean   1930
Egg-ended boilers, makers and date unknown, about 4ft 6in x 35ft long. 40lb/in².

Lightmoor was probably the last colliery to be steamed entirely by egg-ended boilers, with a dozen or so on the surface and two down the pit. It was a well kept colliery, with full brick casing over the boiler tops, and an absence of leaks, or the wooden plugs often driven into leaky joints in the steam mains. The steam and water pipes were all straight runs, with slip expansion joints between each boiler. The pipes, flanges and fittings were all of cast iron with curved faces to fit directly on to the boiler shells, so avoiding separate pads rivetted to the boiler for the fittings. The water-level indicators were swinging levers with arch heads for the chains, and the nearer two show the boilers to be full of water ready for steaming. The smaller pipe was for the feed water, delivering to the boiler top and an internal pipe to the back of the boiler. The steam main above had its own safety valve, and the slip joints had check rods to prevent them from blowing out. The fittings were the simplest possible castings readily made by the local foundry. Conditions everywhere were a tribute to the enginewrights — there simply were no leaks, even on the main winding engine steam range at full boiler pressure, with six boilers and a dozen sliding joints.

120   Lightmoor Colliery. Fronts and firing aisle.

This shows the simple practical nature of the egg-ended boiler in collieries. The wall that took the weight of the supporting brackets fixed to the boiler shell, was brought forward, with circular supporting plates and independent furnace fronts. It was not easy to see the water level indicators from the firing floor, but the great volume of water they contained made such checks an infrequent chore.

18

19

121   Near Staveley, Derbyshire. Rastrick boiler, took waste heat from four boilers.   1960
Maker and date unknown, 11ft diameter x 20ft high. 30lb/in².
Rastrick boilers were usually 2 to 2½ diameters high, and were set vertically. The gases from the furnaces impinged upon the outer shell at the bottom, travelling up the shell and through the radial flues into the large central down flue to the chimney. One radial flue is seen near the top left, and the next, from the inside of the downflue. The heat from the furnaces varied from full flame to cool non-oxydising smoke during the two hour puddling cycle, and with four furnaces at work the irregular heating tried them severely. The large patches on the lower curve were due to the intense flame heat.

122   Whitwick Colliery Co, Leicestershire, Swannington Pumping Station.   1948
Five Cornish boilers, R. Stephenson and Co, Newcastle-upon-Tyne, 1877. About 26ft x 7ft. Furnace flues 3ft 6in diameter. 60lb/in².
Introduced by Richard Trevithick early in the nineteenth century, internally fired boilers became standard. They had many advantages, ie no intense heating of the shell, the fire heat was transferred directly to the water, and the plating was simple, of flat rolled plates and angle iron. The grate area was limited by the diameter of the flue, and grates longer than 7ft were difficult to fire. These were almost certainly the original boilers supplied by Stephensons with the engine (Plate 51) and were still insured for the original pressure at 70 years old. Structurally weak, they were supported on moulded seating blocks which provided a gas circuit along the sides and under the shell on the way to the chimney. Powerful electric pumps had to be installed following an inundation, and were retained when the steam plant was scrapped.

List of Engine Builders

D. Adamson, 98
A. Barclay & Co, 15, 60
Bellis & Morcom, 113
Bever, Dorling & Co, 26
Broseley Foundry, *Frontis*
Butterley Co, 34
Butterworth, 110
Cochrane, Grove & Co, 76
Cole, Marchent & Morley, 95, 102, 106
Daglish & Co, 14
J. Davies, 116
Davy Bros, 78, 84
Dunston Engine Co, 3
W. Evans, 18
E. Foden, 96, 97
J. Fowler, 22
Fraser & Chalmers, 30
Fullerton, Hodgart & Barclay, 108
Galloways, 69, 70-1, 81, 92-3, 99-101
Goscote Foundry, 40
Grant & Ritchie, 25
Greenhalgh & Co, 24
A. Handyside, 19
Harris, 65(?)
Harvey & Co, 44
Hathorn Davey, 49, 50
Hick Hargreaves, 91, 94, 103-4
Holcroft (?), 88

Holman, 45
J. Joicey, 16
Kitson & Co, 72-3
Lamberton, 82-3, 107
Lilleshall, 1(?), 10(?)
C. Markham, 80, 112
Morgan Engineering Co, 117
Thomas Murray, 4-6, 8-9, 61-2
J. Musgrave, 13, 28, 29, 56, 90, 115
Nasmyth Wilson, 11, 20
Newton Chambers, 41-2
Perran Foundry, 38
Perry(?), 88
Thomas Richardson, 7
Richardson Westgarth, 74-5, 77
Robey & Co, 27
Sandys Vivian, 46-8
Scott & Hodgson, 79, 85
Sheepbridge, 86
R. Stephenson & Co, 51-2, 122
J. Stevenson & Co, 23
Summers & Scott, 105
Thornewill & Warham, 32, 36, 39
J. R. Waddell, 53-5
Walker Bros, 57-8
Worsley Mesnes Co, 21(?)
Yates & Thom, 31, 33
Unknown, 2, 17, 37, 40, 59, 63-8, 89

List of Engines Described

Mine Winding Engines

Beam Engines
Frontis	Broseley Tile Works	Broseley, Shropshire	(Broseley Foundry?)
1	Prior Lee Pits	St. George's, Shropshire	(The Lilleshall Co)
2	Sevens Pit	Walsall	(Unknown)
3	Oswald Pit	Lanchester, Co Durham	(Dunston Engine Co)

Vertical Single-Cylinder Engines
4-6	Fortune Pit	Burnhope Colliery	(Thos Murray & Co)
7	South Hetton Colliery	Lambton & Hetton Collieries	(Thos Richardson)
8-9	Wearmouth A & B Pits	Sunderland	(Thos Murray & Co)

Vertical Twin-Cylinder Engines
10	Rockingham Colliery	Staveley Ironworks	(The Lilleshall Co)
11-12	Sandholes Colliery	Manchester Collieries	(Nasmyth Wilson)
13	Wombwell Main Colliery	Wombwell	(J. Musgrave & Son)

Inverted Vertical Twin-Cylinder Engines
14	Tirpentwys Colliery	Pontnewynydd	(Daglish & Co)
15	Elliott Colliery	Elliott, S. Wales	(A. Barclay & Co)

Horizontal Single-Cylinder Engines
16	Burnhope Colliery	Burnhope, Co Durham	(J. Joicey & Co)
17	Montcliff Colliery	Adam Mason and Son	(Unknown)

Horizontal Twin-Cylinder, Slide Valve, Engines
18	Old Mills Colliery	Radstock, Somerset	(Wm. Evans)
19	Morton Colliery	The Clay CRoss Co	(A. Handyside)
20	Cwm Colliery	The Ebbw Vale Co	(Nasmyth Wilson)
21	Nailstone Colliery	Ellistown, Leics	(Worsley Mesnes?)

Horizontal Twin-Cylinder, Cornish & Drop Valve Engines
22	Abercynon Colliery	Glamorgan	(J. Fowler & Co)
23	Wood Pit	Tyldesley, Lancs	(Stevenson & Co)
24	Chanters Colliery	Atherton, Lancs	(Greenhalgh & Co)
25	Woodhorn Colliery	Ashington, Northumberland	(Grant & Ritchie)
26	Wearmouth Colliery	Sunderland	(Bever, Dorling & Co)
27	South Celynen Colliery	South Wales	(Robey & Co)
28	Brackley Colliery	Little Hulton, Lancs	(J. Musgrave & Son)
29	Nook Colliery	Astley, Lancs	(J. Musgrave & Son)

Other Types
30	Penallta Colliery	South Wales	(Fraser & Chalmers)
31	Astley Green Colliery	Astley, Lancs	(Yates & Thom)
32	Elliott Colliery	New Tredegar	(Thornewill & Warham)
33	Sutton Manor Colliery	St Helens	(Yates & Thom)

Angle Compound Engines
34, 35	Cinderhill Colliery	Notts	(Butterley Co)
36	Bargoed Colliery	Powell Duffryn Co	(Thornewill & Warham)

Mine Pumping Engines

Beam Engines
37	Bagworth Colliery	Leics	(Unknown)
38	Hodbarrow Iron Mine	Cumberland	(Perran Foundry)
39	Mill Close Lead Mine	Darley Dale, Derbys	(Thornewill & Warham)
40	Griff Colliery	Nuneaton	(Goscote Foundry)
41, 42	Newbegin Shaft	Newton Chambers Co	(Newton Chambers?)

Beam Engines in Cornwall
44	East Pool	Camborne	(Harvey)
45	East Pool	Camborne	(Holman)
46, 47	Parkandillick Clay Works	St Austell	(Sandys Vivian)
48	South Crofty	Camborne	(Sandys Vivian)

Davey Differential Engine
49	Wood End Pit	Bank Hall Colliery, Burnley	(Hathorn Davey)
50	Clifton Colliery	Hargreaves, Burnley	(Hathorn Davey)

Tandem—Compound Rotative Engine
51, 52	Whitwick Collieries	The Calcutta Engine	(R. Stephenson & Co)

Mine Ventilation Engine

Waddell Fan Engines
53, 54	Morton Colliery	Clay Cross Co	(J. R. Waddell)
55, 56	Chanters Colliery	Tyldesley, Lancs	(J. Musgrave & Son)

Walker Fan Engines
57	Crumlin Navigation Colliery	S Wales	(Walker Bros)
58	Dean & Chapter Colliery	Co Durham	(Walker Bros)

Drift Mine and Haulage Engines

59	Smeaton Mine (Drift)	Lothian	(Unknown)
60	Morlais Colliery (Drift)	Llanelly	(A. Barclay & Co)
61	Warden Law (Bank Haulage)	Hetton Colliery	(Thos Murray & Co)
62	Bank Top (Bank Haulage)	Burnhope Colliery	(Thos Murray & Co)
63	A South Wales Colliery		(Unknown)
64	Lightmoor Colliery	Forest of Dean	(Unknown)

Blast Furnace Blowing Engines

Beam Engines
65	Westbury Ironworks	Wilts	(Unknown)
66, 67	G. & R. Thomas	Hatherton, Staffs	(Unknown)
68	Goldendale Ironworks	Stoke-on-Trent	(Unknown)

Other Blowing Engines

69	Westbury Ironworks	Wilts	(Galloways)
70	Staveley Ironworks	Derbys	(Galloways)
71	Kettering Ironworks	Northants	(Galloways)
72, 73	Kettering Ironworks	Northants	(Kitson)
74	Briton Ferry Ironworks	South Wales	(Richardson Westgarth)
75	Appleby Frodingham Ironworks	Scunthorpe	(Richardson Westgarth)
76	Gjers Mills & Co	Middlesbrough	(Cochrane, Grove & Co)
77	Brymbo Iron and Steel Works	North Wales	(Richardson Westgarth)

Rolling Mill Engines

Reversing Rolling Mill Engines

78	Park Gate Iron & Steel Co	Cogging Mill	(Davy Bros)
79	Park Gate Iron & Steel Co	Cogging Mill (later engine)	(Scott & Hodgson)
80	Park Gate Iron & Steel Co	Cogging Mill (roughing mill)	(C. Markham & Co)
81	Park Gate Iron & Steel Co	Finishing Mill	(Galloways)
82	Park Gate Iron & Steel Co	Billet Mill	(Lamberton)
83	Bynea Steel Works	Loughor	(Lamberton)
84	English Steel Corp	River Don Works	(Davy Bros)
85	John Summers & Co	Shotton Steel Works	(Scott & Hodgson)
86, 87	Sheepbridge Coal & Iron Co		(Sheepbridge Works)

Midlands Ironworks Engines

88	Harts Hill Ironworks	Dudley	(Holcroft?)
89	Walkers' Rolling Mills	Walsall	(Unknown)

Continuous and Sheet Rolling Mill Engines

90	Glynhir Tinplate Co	Pontardulais	(J. Musgrave & Son)
91	Gorse Galvanising Co	Llanelly	(Hick Hargreaves)
92, 93	J. Lysaght Ltd	Newport	(Galloways)
94	Partridge Jones & John Paton	Pontnewynydd	(Hick Hargreaves)
95	Pontardawe Alloy Co	Pontardawe	(Cole, Marchent & Morley)

Tinplate Hot-Mill Engines

96	Kidwelly Tinplate Co	Kidwelly	(Edwin Foden)
97	Baglan Bay Tinplate Co	Briton Ferry	(E. Foden & Co)
98	St David's Tinplate Co	Loughor	(D. Adamson)
99	Bryngwyn Sheet Works	Gorseinon	(Galloways)
100	Baglan Bay Tinplate Works	Briton Ferry	(Galloways)

Tinplate Cold-Rolling Mill Engines

101	Baglan Bay Tinplate Works	Briton Ferry	(Galloways)

102	St David's Tinplate Works	Loughor	(Cole Marchent & Morley)
103	Abercarn Works	Abercarn	(Hick Hargreaves)
104	John Player & Co	Clydach	(Hick Hargraves)
105	Mclingriffith Tinplate Co	Whitchurch	(Summers & Scott)

Engines for Metal Working—General Processes

106	Stewarts & Lloyds	Newport	(Cole, Marchent & Morley)
107	Taylor Bros	Trafford Park	(Lamberton)
108, 109	Thos Whitehead & Co	Scunthorpe	(Fullerton, Hodgart & Barclay)
110, 111	Winterbottoms	Oxspring Wire Mill	(Butterworth)

Works Services

112	Park Gate Iron & Steel Co	Rotherham	(C. Markham & Co)
113	Park Gate Iron & Steel Co	Rotherham	(Belliss & Morcom)
114	Park Gate Iron & Steel Co	Rotherham	(Various)
115	Abercarn Tinplate Works	Abercarn	(J. Musgrave & Son)
116	John Bagnall & Co	Wednesbury	(James Davies)
117	Taylor Bros	Trafford Park	(Morgan Engineering Co)

Boilers

118	Froglane Colliery	Westerleigh, Glos	(Unknown)
119, 120	Lightmoor Colliery	Forest of Dean	(Unknown)
121		Staveley, Derbys	(Unknown)
122	Whitwick Colliery	Swannington, Leics	(R. Stephenson)